**ACPL ITEM
DISCARDED**

INTERNATIONAL SERIES OF MONOGRAPHS IN
SOLID STATE PHYSICS
General Editors: R. Smoluchowski and N. Kurti

Volume 5

STRUCTURE, PROPERTIES AND PREPARATION OF PEROVSKITE-TYPE COMPOUNDS

STRUCTURE, PROPERTIES AND PREPARATION OF PEROVSKITE-TYPE COMPOUNDS

BY

FRANCIS S. GALASSO

United Aircraft Research Laboratories

PERGAMON PRESS
OXFORD · LONDON · EDINBURGH · NEW YORK
TORONTO · SYDNEY · PARIS · BRAUNSCHWEIG

Pergamon Press Ltd., Headington Hill Hall, Oxford
4 & 5 Fitzroy Square, London W.1
Pergamon Press (Scotland) Ltd., 2 & 3 Teviot Place, Edinburgh 1
Pergamon Press Inc., Maxwell House, Fairview Park, Elmsford
New York 10523
Pergamon of Canada Ltd., 207 Queen's Quay West, Toronto 1
Pergamon Press (Aust.) Pty. Ltd., 19a Boundary Street,
Rushcutters Bay, N.S.W. 2011, Australia
Pergamon Press S.A.R.L., 24 rue des Écoles, Paris 5ᵉ
Vieweg & Sohn GmbH, Burgplatz 1, Braunschweig

Copyright © 1969
Pergamon Press Inc.

First edition 1969

Library of Congress Catalog Card No. 68-21881

PRINTED IN HUNGARY

08 012744 4

ERRATA

STRUCTURE, PROPERTIES AND PREPARATION OF PEROVSKITE TYPE COMPOUNDS

by

Francis S. Galasso

Published by

Pergamon Press 1969

p. vi, last line: *for* Powdres *read* Powders

p. 9, line 5: *for* $z + 1/$ *read* $z + 1/2$

p. 69, under Fig. 4.8: *for* Van Stanten *read* Van Santen

p. 157, last three compounds, *for* $O_{3.75}$ *read* $O_{2.75}$

p. 181, ref. 56: *for* Finchbaugh *read* Flinchbaugh

p. 187, last paragraph should read :-

The electrical resistivity of Mn_3ZnC is 770×10^6 ohm-cm at $20°$ and decreases with increasing temperature. It is not stable in moist air and must be kept in a desiccator.

CONTENTS

PREFACE	ix
1. INTRODUCTION	1
2. STRUCTURE OF PEROVSKITE-TYPE COMPOUNDS	3
2.1. Ternary Oxides	4
2.2. Complex Oxides	11
2.3. Madelung Energy	39
2.4. Ionic Radii	41
3. X-RAY DIFFRACTION AND ELECTRON PARAMAGNETIC STUDIES	50
3.1. X-ray Diffraction	50
3.2. Electron Paramagnetic Resonance Studies	57
4. CONDUCTIVITY	60
4.1. Conductors	60
4.2. Superconductors	63
4.3. Semiconductors	64
4.4. Thermoelectricity	73
4.5. Hall Effect	76
5. FERROELECTRICITY	79
5.1. Ternary Perovskites	80
5.2. Solid Solutions	90

CONTENTS

5.3. Complex Perovskites	99
5.4. Effect of Nuclear Irradiation	103
5.5. Applications of Ferroelectric Materials	105
5.6. Theories of Ferroelectricity	109
6. PHASE TRANSITIONS	115
6.1. Ternary Perovskites	115
6.2. Complex Perovskite-type Compounds	118
7. FERROMAGNETISM	122
8. OPTICAL PROPERTIES	129
8.1. Transmittance	129
8.2. Coloration by Light	133
8.3. Electro-optic Effect	133
8.4. Lasers	134
9. OTHER PROPERTIES	140
9.1. Catalysts	140
9.2. Thermal Conductivity	141
9.3. Melting Points	141
9.4. Heats of Formation	142
9.5. Thermal Expansion	143
9.6. Density	143
9.7. Mechanical Properties	144
10. PREPARATION OF PEROVSKITE-TYPE OXIDES	159
10.1. Powdres	159

10.2. Thin Films	162
10.3. Single Crystals	165
11. OTHER PEROVSKITE-TYPE COMPOUNDS	**182**
11.1. Preparation of Perovskite-type Phases	182
11.2. Structure	184
11.3. Properties	187
INDEX	191
OTHER TITLES IN THE SERIES	209

PREFACE

SINCE 1945, when the ferroelectric properties of barium titanate were reported by von Hippel in the United States and independently by workers in other countries, ABO_3 compounds with the perovskite structure have been studied extensively. These studies have resulted in the discovery of many new ferroelectric and piezoelectric materials. Most of the literature written on perovskite-type compounds has been concentrated on these properties.

In addition, a number of solid-state chemists devoted many years to producing new ternary perovskite compounds of all kinds and studying their structures. By 1955 it appeared that most of the possible combinations of large A cations and smaller B ions needed to form perovskite-type compounds had been tried. At that time, as part of a thesis problem at the University of Connecticut, I found that new perovskite-type compounds could be prepared by introducing more than one element in the B position of the perovskite structure.

At the United Aircraft Research Laboratories in 1960, J. (Pyle) Pinto, W. Darby and I continued this research by initiating an extensive program to study the preparation, structure and properties of these perovskite-type compounds. Because these compounds contained two different B ions with different valence states, many combinations of elements and, therefore, the formation of many compounds were possible. These studies as well as research conducted by other workers throughout the world resulted in large amounts of new structural and property data on perovskite-type compounds. During the same time renewed interest in ternary perovskite compounds was generated as a result of studies which showed that they might prove useful as laser host materials, for laser modulation, as thermistors, as superconductors and as infrared windows.

The purpose of this book is to attempt to bring together the information obtained from these studies, including the various methods of preparing powders, thin films and single crystals of perovskite-type compounds, the structure of these compounds and their properties. The properties covered are electrical conductivity, ferroelectricity, ferromagnetism, optical transmittance the electro-optical effect, catalytic properties, melting points, heats of formation, thermal expansion, densities and mechanical properties. Because of the growing number of applications for perovskite-type compounds, I felt that this information might prove valuable to applied researchers. In addition, structural data are included for scientists who are interested in correlating the structure and properties of materials.

I am grateful to Professor Roland Ward and Professor Lewis Katz for introducing me to this field of research and to my previous fellow workers at the University of Connecticut for their studies on many unusual perovskite-type compounds. I must also acknowledge Professor Aaron Wold and Professor Rustum Roy for their discussions on perovskite compound preparations, Dr. Michael Kestigian and Professor A. Smakula for helpful advice on crystal growing, Dr. Alexander Wells and Professor Martin Buerger for pointing out the need for a compilation and discussion of structural data of the type presented herein, Dr. Fredrick Seitz for helpful discussions on ordering and to Dr. John Goodenough for information on the ferromagnetic properties and conductivity in perovskites. I would also like to thank Dr. V. Nicolai of O.N.R. Washington, D.C., for information on lasers, Dr. Charleton of Fort Monmouth, New Jersey, for discussions on dielectrics and Dr. Fredricks of Wright-Patterson A.F.B. for reports on microwave properties of perovskites. I am indebted to United Aircraft Research Laboratories, my colleagues J. Pinto and W. Darby; R. Fanti, Chief of Materials Sciences, and Professor P. Duwez, a member of the Advisory Committee for United Aircraft Corporation. I wish to thank Professor R. Smoluchowski, Dr. R. Graf, Professor A. Wold and Dr. M. Kestigian for helpful suggestions and for checking through the manuscript. Finally, I am grateful to my wife, Lois, Miss Kathy Donahue, Miss Joyce Hurlburt, Mrs. Jean Kelly and Mrs. Nancy Letendre for their patience and effort in preparing this manuscript.

CHAPTER 1

INTRODUCTION

THIS book contains details on the structure, properties, and preparation of perovskite-type compounds. Because of the growing number of applications for these compounds, information on their preparation is becoming more in demand. The long fluorescence lifetimes observed for Cr^{3+} in $LaAlO_3$, and the large room-temperature electro-optical effect in $K(Ta_{0.65}Nb_{0.35})O_3$, for example, have caused considerable interest in obtaining these materials as optical quality single crystals. In addition, the better known ferroelectric and piezoelectric properties of perovskites have induced researchers to continue the effort to prepare them as larger and more perfect single crystals, polycrystalline compacts and thin films. Materials scientists also are continuously trying to prepare new perovskite compounds with new and improved properties. One of the best ways of accomplishing this is to use the insight gained from structure–property relationships. An objective of this book is to point out some of these structure–property relationships as well as to provide the reader with enough data so that he can deduce some of his own.

In this book, the oxide phases have been divided into two types, the ternary ABO_3 type and their solid solutions, where A is a large metal cation and B is a smaller metal cation and the newer complex $A(B'_xB''_y)O_3$ type compounds where B' and B'' are two different elements in different oxidation states and $x+y = 1$. First, the structural data are presented in a systematic manner for quick and easy reference. A chapter is included on the identification of distortions in the structure of ternary perovskite-type compounds and of ordering in the structure of complex perovskite-type compounds using X-ray diffraction techniques. The properties of the perovskite compounds described herein are electrical conductivity, ferroelectricity, ferromagnetism, optical, catalytic, melting points, heats of formation, thermal expansion and mechanical properties. Then, the preparation of these compounds as powders, thin films and single crystals are described. In addition, a chapter is included on other compounds besides oxides with the perovskite structure.

CHAPTER 2

STRUCTURE OF PEROVSKITE-TYPE COMPOUNDS

Most of the compounds with the general formula ABO_3 have the perovskite structure. The atomic arrangement in this structure was first found for the mineral perovskite, $CaTiO_3$. It was thought that the unit cell of $CaTiO_3$ could be represented by calcium ions at the corners of a cube with titanium ions at the body center and oxygen ions at the center of the faces (Fig. 2.1). This simple cubic structure has retained the name perovskite, even though $CaTiO_3$ was later determined to be orthorhombic by Megaw.[1] Through the years it has been found that very few perovskite-type oxides

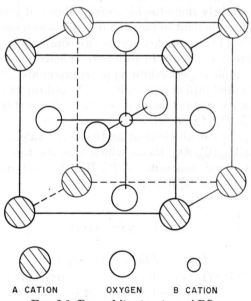

FIG. 2.1. Perovskite structure, ABO_3.

have the simple cubic structure at room temperature, but many assume this ideal structure at higher temperatures.

In the perovskite structure, the A cation is coordinated with twelve oxygen ions and the B cation with six. Thus, the A cation is normally found to be somewhat larger than the B cation. In order to have contact between the A, B, and O ions, $R_A + R_O$ should equal $\sqrt{2}(R_B + R_O)$, where R_A, R_B and R_O are the ionic radii. Goldschmidt[2] has shown that the cubic perovskite structure is stable only if a tolerance factor, t defined by $R_A + R_O = t\sqrt{2}(R_B + R_O)$, has an approximate range of $0.8 < t < 0.9$, and a somewhat larger range for distorted perovskite structures. It should be noted that conflicting reports in the literature make it difficult to assign the correct unit cell dimensions for these distorted perovskite structures.

The ternary perovskite-type oxides described in this chapter will be divided into $A^{1+}B^{5+}O_3$, $A^{2+}B^{4+}O_3$, $A^{3+}B^{3+}O_3$ types and oxygen- and cation-deficient phases. The oxygen- and cation-deficient phases will be regarded as those which contain considerable vacancies and not those phases which are only slightly non-stoichiometric. Many of these contain B ions of one element in two valence states and should not be confused with the complex perovskite compounds which contain different elements in different valence states.

The complex perovskite type compounds, $A(B'_xB''_y)O_3$, will be divided into compounds which contain twice as much lower valence state element as higher valence state element, $A(B'_{0.67}B''_{0.33})O_3$, those which contain twice as much of the higher valence state element as the lower valence state element, $A(B'_{0.33}B''_{0.67})O_3$, those which contain the two B elements in equal amounts, $A(B'_{0.5}B''_{0.5})O_3$, and oxygen-deficient phases $A(B'_xB''_y)O_{3-z}$.

2.1. TERNARY OXIDES

Oxides of the $A^{1+}B^{5+}O_3$ Type

The $A^{1+}B^{5+}O_3$ type oxides are of particular interest because of their ferroelectric properties. Potassium niobate, $KNbO_3$ has a structure which can be described by an orthorhombic unit cell of $a = 3.9714$ Å, $b = 5.6946$ Å and $c = 5.7203$ Å where b and c equal approximately $\sqrt{2}\,a$ or the length of face diagonals of the simple perovskite cell, and exhibits ferroelectric properties. The sodium niobate, $NaNbO_3$, structure also can be described by an orthorhombic unit cell but is antiferroelectric. The unit cell is pseudotetragonal at 420°C, tetragonal at 560°C and cubic at 640°C. Unlike the structure of niobates the $KTaO_3$ structure is described by a cubic unit cell. The structure of $NaTaO_3$ is orthorhombic with the space group $Pc2_1n$ and all the atoms are placed in the unit cell in positions:[3]

(4a) $xyz;\ \bar{x}, y+\tfrac{1}{2}, \bar{z};\ x+\tfrac{1}{2}, y+\tfrac{1}{2}, \tfrac{1}{2}-z;\ \tfrac{1}{2}-x, y, z+\tfrac{1}{2}$

	x	y	z
Na	−0.01	0.78	0.02
Ta	0.50	0.00	0.00
O(1)	−0.02	0.76	0.52
O(2)	0.29	−0.03	0.29
O(3)	0.29	0.56	0.29

Smith and Welch[4] found that potassium iodate, KIO_3, and thallous iodate, $TlIO_3$, also adopt the perovskite structure. Single-crystal studies showed that KIO_3 had a rhombohedral structure with unit cell parameters $a = 4.410$ Å, $\alpha = 89.41°$. Powder diffraction studies on $TlIO_3$ powders indicated that it also had a rhombohedral structure with cell dimensions, $a = 4.510$ Å, $\alpha = 89.34°$, while the structures of $CsIO_3$ and $RbIO_3$, on the other hand, have been reported to be cubic.

Oxides of the $A^{2+}B^{4+}O_3$ Type

Probably the largest number of perovskite-type compounds are described by the general formula $A^{2+}B^{4+}O_3$, where the A cations are alkaline earth ions, cadmium or lead and the B^{4+} ions include Ce, Fe, Pr, Pu, Sn, Th, Hf, Ti, Zr, Mo and U. The best known compounds of this type are the titanates because of the ferroelectric properties that the barium and lead compounds exhibit. Calcium titanate, as previously mentioned, was the original example of a compound with an "ideal" cubic perovskite structure, but it was later determined to have an orthorhombic structure. The structure of strontium titanate, however, is truly cubic; the space group is $Pm3m$, and its atoms are in the following positions in the unit cell:

Sr: 0 0 0

Ti: $\frac{1}{2}$ $\frac{1}{2}$ $\frac{1}{2}$

O: $\frac{1}{2}$ $\frac{1}{2}$ 0; $\frac{1}{2}$ 0 $\frac{1}{2}$; 0 $\frac{1}{2}$ $\frac{1}{2}$

Barium titanate and lead titanate are of more interest because the atomic displacements in their structures produce ferroelectric properties. Neutron diffraction studies show that the displacements are greater in lead titanate than they are in barium titanate. For barium titanate the atoms in the unit cell are in the following positions:

Ba atom at 0 0 0

Ti atom at $\frac{1}{2}$, $\frac{1}{2}$, 0.512   (ref. 3)

one O atom at $\frac{1}{2}$, $\frac{1}{2}$, 0.023

and two O atoms at $\frac{1}{2}$, 0, 0.486; 0, $\frac{1}{2}$, 0.486.

The room temperature tetragonal form of lead titanate has its atoms in these positions in its unit cell:

Pb atom at 0 0 0

Ti atom at $\frac{1}{2}$, $\frac{1}{2}$, 0.541   (ref. 3)

one O atom at $\frac{1}{2}$, $\frac{1}{2}$, 0.112

and two O atoms at $\frac{1}{2}$, 0, 0.612; 0, $\frac{1}{2}$, 0.612.

While the structure of calcium titanate exhibits orthorhombic symmetry at room temperature, it becomes cubic above 900°C. Barium titanate undergoes the following transformations:

Rhombohedral $\xrightarrow{-100°C}$ orthorhombic $\xrightarrow{0°C}$ tetragonal $\xrightarrow{120°C}$ cubic and lead titanate transforms tetragonal $\xrightarrow{490°C}$ cubic.

Roth[5] regards $BaZrO_3$ as another compound with an "ideal" cubic perovskite structure. While his conclusion has been questioned, it is logical that this compound with a tolerance factor of 0.88 should adopt the same structure as $SrTiO_3$ which has a tolerance factor of 0.86. Strontium zirconate and calcium zirconate probably have an orthorhombic structure, although Smith and Welch[4] felt that the $SrZrO_3$ powder pattern should be indexed on a "doubled" cubic perovskite

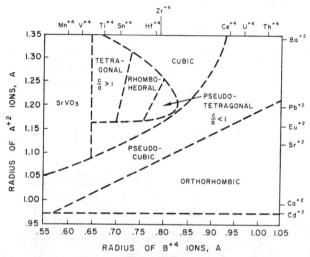

Fig. 2.2. Classification of perovskite-type compounds, $A^{2+}B^{4+}O_3$ (after Roth[5]).

cell. The lead zirconate structure originally was thought to be tetragonal, but was found to be orthorhombic by X-ray and neutron diffraction studies.

The compound $BaSnO_3$ also has been reported by Smith and Welch[4] to have the "ideal" cubic perovskite structure. This selection of the unit cell was confirmed by Roth[5] and Megaw[1] independently.

Another interesting compound is $CaUO_3$, because although it has a tolerance factor of only 0.71 it still has the perovskite structure. Roth[5] points out that a minimum tolerance factor of 0.77 was previously set for $A^{2+}B^{4+}O_3$ type compounds, but because of this new information can be assumed to be incorrect. The compound $CaUO_3$ was not found to have a cubic structure but was found to adopt the $CaTiO_3$ structure.

A diagrammatic presentation of radius data for $A^{2+}B^{4+}O_3$ type compounds is shown in Fig. 2.2. The regions are determined from experimental data for room-temperature studies. While this diagram holds well for compounds, there are some discrepancies in the boundaries of the ferroelectric field for solid solutions. The diagram, however, is still a useful summary of structural data.

Oxides of the $A^{3+}B^{3+}O_3$ Type

The largest number of $A^{3+}B^{3+}O_3$ type compounds were found by Geller and Wood[6] to have an orthorhombic structure similar to that for GdFeO$_3$, Fig. 2.3. The space group for these compounds is *Pbnm* and the atoms are in the following positions:

Four Gd atoms at $\pm(x, y, \frac{1}{4}; \frac{1}{2}-x, y+\frac{1}{2}, \frac{1}{4})$ $x = -0.018$
$y = 0.060$.

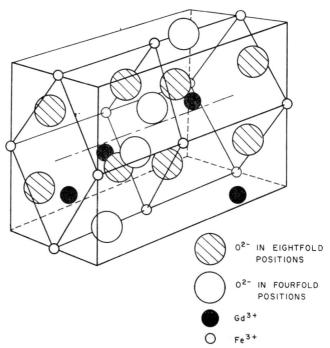

Fig. 2.3. Structure of GdFeO$_3$ (after Geller[6]).

Four Fe atoms at $\frac{1}{2}$ 0 0; $\frac{1}{2}$ 0 $\frac{1}{2}$; 0 $\frac{1}{2}$ 0; 0 $\frac{1}{2}$ $\frac{1}{2}$.

Four O atoms at the same positions listed for Gd atoms, but with $x = 0.05$, $y = 0.47$ (ref. 3)

and

eight O atoms at $\pm(xyz;\ \frac{1}{2}-x,\ y+\frac{1}{2},\ \frac{1}{2}-z;\ \bar{x},\ \bar{y},\ z+\frac{1}{2};\ x+\frac{1}{2},\ \frac{1}{2}-y,\ \bar{z}$

where $x = -0.29$, $y = 0.275$, $z = 0.05$.

The relationship of the orthorhombic unit cell to that of the perovskite structure can be seen in Fig. 2.3. The unit cell for the GdFeO$_3$ structure, $a_0 = 5.346$ Å, $b_0 = 5.616$ Å, $c = 7.668$ Å contains four distorted perovskite units.

Looby and Katz[7] thought they had found a new type of structure adopted by YCrO$_3$, and indexed the powder pattern on the basis of a monoclinic cell, but pointed out that the correct unit cell might be orthorhombic. Geller and Wood[6] were able to show from single crystal studies that the structure of YCrO$_3$ was similar to that of GdFeO$_3$. Thus, it is quite possible that many additional $A^{3+}B^{3+}O_3$ compounds which have been reported to have monoclinic structures could really have orthorhombic structures. Some of the compounds which were confirmed as having the GdFeO$_3$ structure are EuAlO$_3$, EuFeO$_3$, GdAlO$_3$, GdCrO$_3$, GdVO$_3$, LaAlO$_3$, LaCrO$_3$, LaGaO$_3$, LaScO$_3$, NdAlO$_3$, NdCrO$_3$, NdFeO$_3$, NdVO$_3$, NdGaO$_3$, PrCrO$_3$, PrFeO$_3$, PrGaO$_3$, PrScO$_3$, PrVO$_3$, SmAlO$_3$, SmCrO$_3$, SmFeO$_3$, YScO$_3$, YAlO$_3$, YCrO$_3$, YFeO$_3$, NdScO$_3$ and GdScO$_3$. Two of these compounds, LaGaO$_3$ and SmAlO$_3$ transform to a rhombohedral form at 900° and 850°C respectively. The compounds LaAlO$_3$, NdAlO$_3$ and PrAlO$_3$ also have this rhombohedral structure which is probably quite similar to that of GdFeO$_3$.

Figure 2.4 presents a classification of $A^{3+}B^{3+}O_3$ type compounds according to the constituent ionic radii. All of the compounds in the upper left of the diagram form perovskite-type structures. Where both the A and B ions are small, the compounds have the corundum- or ilmenite-type structures. When both the A and B ions are large, the phases form La$_2$O$_3$ type structures.

While none of the $A^{3+}B^{3+}O_3$ type compounds have the "ideal" cubic perovskite structure, the rhombohedral perovskites such as $LaAlO_3$ are only slightly distorted. The search for laser host materials with cubic crystallographic sites for Cr^{3+} substitution has produced considerable interest in these compounds. Lanthanum aluminum oxide, $LaAlO_3$, with $\alpha = 90°4'$, has been widely studied as a laser host material. However, the phase transition at $435°$ has presented considerable problems in trying to grow it in single crystal form.

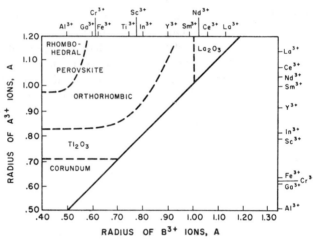

FIG. 2.4. Classification of perovskite-type compounds, $A^{3+}B^{3+}O_3$ (after Roth[5]).

Non-stoichiometric Ternary Oxides

Probably the best known non-stoichiometric ternary oxides are the tungsten bronzes. The phases Na_xWO_3 have been found to have the cubic perovskite structure in the range $0.3 < x < 0.95$[8] and the phases Li_xWO_3 in the range $0.35 < x < 0.57$.[8] The lattice constants of these materials vary linearly with increasing amounts of alkali metal ion. A smaller range of non-stoichiometry exists in the strontium niobium bronzes[9] where the alkaline earth metal ion mole fraction can vary from 0.7 to 0.9 and in La_xVO_3 where $0.66 < x < 1$.

Rooksby et al.[10] reported the preparation of a group of perovskite-type rare earth niobates and tantalates. The structure of these $A_{0.33}BO_3$ type compounds was tetragonal, orthorhombic or monoclinic.

The existence of these cation deficient compounds is not surprising in view of the fact that ReO_3 is stable without A ions. The deficiencies can be tolerated over ranges of composition without changes in structure. However, different amounts of A ion are necessary to stabilize the structure depending on which B ion is in the octahedrally coordinated sites.

Oxygen deficiencies have also been observed in the perovskite structure. The phases $SrBO_{3-x}$ where B is Ti or V have been found to have the perovskite structure over the range $0 < x < 0.5$ for the titanium phases and $0 < x < 0.25$ for the vanadium phases. Both $SrVO_{2.75}$ and $SrTiO_{2.5}$ were found to have cubic structures. Similar phases have been reported in the $SrFeO_{3-x}$,[11-13] $CaMnO_{3-x}$[13, 14] and $SrCoO_{3-x}$[13] systems although the oxygen deficiency is not as great.

Coates and McMillan showed that cubic perovskite structures could be obtained in calcium perovskites, which are normally distorted, by producing oxygen vacancies. The phases $CaMnO_3$ and $CaTiO_3$ become cubic with the introduction of deficiencies. As these authors point out, studies in this area are not abundant enough to obtain a good understanding of the effect of nonstoichiometry on the perovskite structure.

2.2. Complex Oxides

Oxides of the $A^{2+}(B^{3+}_{0.67}B^{6+}_{0.33})O_3$ Type

The structure of compounds which contain twice as many B^{3+} ions as B^{6+} ions is not well established. Fresia et al.,[15] who prepared one of the first compounds of this type, $Ba(Sc_{0.67}W_{0.33})O_3$, felt that it probably had an ordered perovskite structure described by Steward and Rooksby.[16] In this structure the two different B ions alternate at the corners of the simple cubic unit cell of the perovskite structure so that the cell edge has to be doubled (see Fig. 2.5). The space group is $Fm3m$ and the atomic positions are:

A: $\frac{1}{4}\frac{1}{4}\frac{1}{4}$; $\frac{1}{4}\frac{3}{4}\frac{3}{4}$; $\frac{3}{4}\frac{1}{4}\frac{3}{4}$; $\frac{3}{4}\frac{3}{4}\frac{1}{4}$; $\frac{3}{4}\frac{3}{4}\frac{3}{4}$; $\frac{1}{4}\frac{1}{4}\frac{3}{4}$; $\frac{1}{4}\frac{3}{4}\frac{1}{4}$; $\frac{3}{4}\frac{1}{4}\frac{1}{4}$

B: $0\ 0\ \tfrac{1}{2}; \tfrac{1}{2}\ 0\ 0; 0\ \tfrac{1}{2}\ 0; \tfrac{1}{2}\ \tfrac{1}{2}\ \tfrac{1}{2}$

B'': $0\ 0\ 0; \tfrac{1}{2}\ \tfrac{1}{2}\ 0; 0\ \tfrac{1}{2}\ \tfrac{1}{2}; \tfrac{1}{2}\ 0\ \tfrac{1}{2}$

O: $u\ 0\ 0$; etc. (24 positions)

Since the B' and B'' have to be present in equal amounts in this structure, $Ba(Sc_{0.67}W_{0.33})O_3$ should probably be written

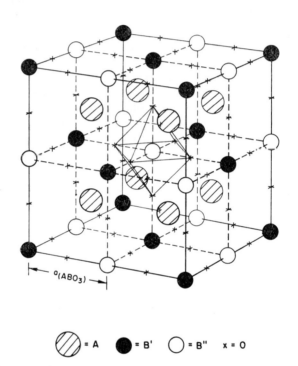

⊘ = A   ● = B'   ○ = B''   x = O

Fig. 2.5. The $(NH_4)_3 FeF_6$ structure: cubic ordered perovskite-type $A(B'_{0.5}B''_{0.5})O_3$ (after L. Pauling, *J. Am. Chem. Soc.* **46**, 2738 (1924)).

$Ba[Sc_{0.5}(Sc_{0.17}W_{0.33})]O_3$ where three-quarters of the Sc ions are on the B' sites and one-quarter of the Sc ions are randomly distributed with the W atoms on the B'' sites. Sleight and Ward[17] also found that it was necessary to use the doubled unit cell to index all observed lines in the X-ray patterns of $A(B'_{0.67}U_{0.33})O_3$ and $A(B'_{0.67}Re_{0.33})O_3$[18] compounds. When the compound $Sr(Cr_{0.5}Re_{0.5})O_3$ was changed in composition to obtain $Sr(Cr_{0.67}Re_{0.33})O_3$ no new phase appeared but the lattice expanded and the intensities of the superlattice lines in the X-ray patterns showed a marked decrease. However, no detailed structure studies have been conducted to determine the arrangement of atoms in these phases.

Oxides of the $A^{2+}(B^{2+}_{0.33}B^{5+}_{0.67})O_3$ Type

A large number of compounds containing niobium and tantalum as one of the B ions in perovskite structure and a divalent ion as the other B ion were prepared by Roy[19] and independently by Galasso et al.[20] Both workers originally could not account for the extra lines which most of the X-ray patterns of these compounds contained. Once Galasso et al.[21] found that one of the compounds, $Ba(Sr_{0.33}Ta_{0.67})O_3$, had a new ordered perovskite structure; subsequent studies showed that many of the $A^{2+}(B^{2+}_{0.33}B^{5+}_{0.67})O_3$ adopted this structure. The structure of $Ba(Sr_{0.33}Ta_{0.67})O_3$ is based on space group $P\bar{3}m1$ with atoms in the unit cell at the following locations:

Ba_1: 0 0 0

Ba_2: $\frac{1}{3}$ $\frac{2}{3}$ z; $\frac{2}{3}$ $\frac{1}{3}$ $\bar{z}$; $z = \frac{2}{3}$

Sr: 0 0 $\frac{1}{2}$

Ta: $\frac{1}{3}$ $\frac{2}{3}$ z; $\frac{2}{3}$ $\frac{1}{3}$ $\bar{z}$; $z = \frac{1}{6}$

O_1: $x, \bar{x}, z$; $x, 2x, z$; $2\bar{x}, \bar{x}, z$;
$\bar{x}, x, \bar{z}$; $\bar{x}, 2\bar{x}, \bar{z}$; $2x, x, \bar{z}$; $x = \frac{1}{6}, z = \frac{1}{3}$

O_2: $\frac{1}{2}$ 0 0; 0 $\frac{1}{2}$ 0; $\frac{1}{2}$ $\frac{1}{2}$ 0

The structure is shown in Figs. 2.6 and 2.7. Note that if the perovskite structure is described as close-packed layers of A and oxygen ions perpendicular to the [111] direction with small B ions in the octahedral holes between the layers, then these B ions Sr and Ta each form planes of atoms. These planes are parallel to the close-packed layers and, since there are twice as many tantalum ions as strontium ions, the repeat scheme is two layers of tantalum ions and one of strontium ions. It is interesting that the ordered structure of $A(B'_{0.5}B''_{0.5})O_3$ type compounds when observed in the same way in the [111]

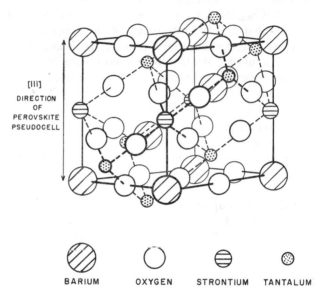

Fig. 2.6. The $Ba(Sr_{0.33}Ta_{0.67})O_3$ structure (after Galasso et al.[21]).

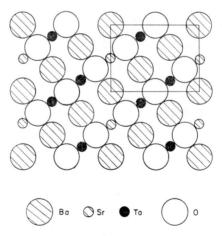

Fig. 2.7. Three-layer repeat sequence in $Ba(Sr_{0.33}Ta_{0.67})O_3$. The (110) plane (after Galasso et al.[21]).

direction, consists of alternating layers, one of which contain B' ions and the other B'' ions.

A study of $Ba(B^{2+}_{0.33}Ta^{5+}_{0.67})O_3$ compounds showed a decrease in the ordering as the difference in the size of the B^{2+} and Ta^{5+} ions became smaller.[22] The results were confirmed in a later investigation of $Ba(B^{2+}_{0.33}Nb^{5+}_{0.67})O_3$ compounds.[23] In some cases the ordering lines in the X-ray patterns were very broad and became sharp when the samples were annealed at high temperatures. This observation was attributed to the existence of small ordering domains which grew at higher temperatures. Thus the X-ray patterns of some of the compounds which might be expected to be ordered because of the large size difference of the B' and B'' ions did not show the ordering lines if they could not be annealed at high temperatures because of low melting points.

Oxides of the $A^{2+}(B^{3+}_{0.5}B^{5+}_{0.5})O_3$, $A^{2+}(B^{2+}_{0.5}B^{6+}_{0.5})O_3$, $A^{2+}(B^{1+}_{0.5}B^{7+}_{0.5})O_3$ and $A^{3+}(B^{2+}_{0.5}B^{4+}_{0.5})O_3$ Types

The largest group of complex perovskite type compounds has the general formula $A(B'_{0.5}B''_{0.5})O_3$. When the structure of these compounds is ordered, and most of them are, they adopt the structure shown in Fig. 2.5. It was postulated by Galasso et al.[20] that an ordered distribution of the B ions is most probable when a large difference existed in either their charges or ionic radii. This hypothesis can be validated qualitatively by looking at the data below in Table 2.1.

TABLE 2.1. *Structural Data for $A(B'_xB''_y)O_3$ Type Compounds*

Compound	Difference in charge of B ions	Difference in ionic radii of B ions	Arrangement of B ions
$Ba(Fe^{3+}_{0.5}Ta^{5+}_{0.5})O_3$	2	0.04	Random[20]
$Ba(Mg^{2+}_{0.5}W^{6+}_{0.5})O_3$	4	0.05	Ordered[16]
$Ba(La^{3+}_{0.5}Ta^{5+}_{0.5})O_3$	2	0.46	Ordered[24]
$Ba(Sr^{2+}_{0.33}Ta^{5+}_{0.67})O_3$	3	0.44	Ordered[21]
$Ba(Zn^{2+}_{0.33}Nb^{5+}_{0.67})O_3$	3	0.05	Random[20]

A study of $Ba(M_{0.5}^{3+}Nb_{0.5}^{5+})O_3$[24] type compounds indicated that the critical percentage difference in ionic radii between B ions which causes ordering lies between 7% and 17%.

The first compounds with this ordered structure were reported by Steward and Rooksby,[16] who found that a number of alkaline earth tungstates and molybdates of the $A(A_{0.5}B''_{0.5})O_3$ type where A is an alkaline earth ion and B'' is Mo or W have this structure. The structures of $Ba(Ca_{0.5}W_{0.5})O_3$ and $Ba(Ca_{0.5}Mo_{0.5})O_3$ were reported to be cubic at room temperature while the $Ba(Sr_{0.5}W_{0.5})O_3$ structure was distorted and became cubic only after heating to 500°C. Fresia et al.[15] found that other ions such as Zn^{2+}, Fe^{2+}, Co^{2+} and Ni^{2+} could be used as the B' ion in the ordered perovskite structure without distortion. However, when compounds were prepared with strontium in the A position the structures were distorted. It should be noted at this point that some authors index the tetragonal and orthorhombic distortion of the cubic ordered unit cell by using the ~ 8 Å edges, while others use an ~ 5.7 Å face diagonal for two axes and ~ 8 Å edge for the third.

Sleight and Ward[17] in a study of $A(B_{0.5}^{2+}U_{0.5})O_3$ type perovskites also found that all of the compounds containing strontium in the A position had distorted structures. The unit cells were pseudomonoclinic, but the powder patterns were indexed on a smaller orthorhombic unit cell. Using a tolerance factor

$$t = R_A + R_0/\sqrt{2}\,[(R_B^{2+} + R_B^{5+})/2 + R_0]$$

they calculated that a number of compounds containing barium in the A position should form the cubic ordered perovskite structure, and observed that they did form this structure.

Some of the most interesting compounds were those containing Mo^{5+}, W^{5+} or Re^{5+} as one of the B ions and another paramagnetic ion as the other because of the ferromagnetic properties they exhibited. These compounds will be discussed in detail in another section.

By adjustment of the oxidation state by valence compensation the compounds $Ba(Li_{0.5}Os_{0.5})O_3$ and $Ba(Na_{0.5}Os_{0.5})O_3$

were prepared.[18] It appears that the osmium has a valence state of 7 in these compounds, a value not reported previously for this element. Similar compounds also were prepared with Re^{7+} as the B″ ion.

Lanthanum also has been introduced in the A position of the perovskite structure with two ions in the B position. The symmetry of the structure of these compounds has not been resolved, but it appears that they adopt some modification of the ordered perovskite structure. Preliminary single crystal studies indicate that they could have ordered cubic unit cells.[26] This would be interesting in light of the fact that there are no $A^{3+}B^{3+}O_3$ type ternary oxides with the simple cubic perovskite structure.

Oxides of the $A^{2+}(B^{1+}_{0.25}B^{5+}_{0.75})O_3$ Type

It was expected, because of the large charge difference in the B ions, that $Ba(Na_{0.25}Ta_{0.75})O_3$ and $Sr(Na_{0.25}Ta_{0.75})O_3$ would be ordered. However, no superstructure lines were observed in the X-ray patterns of these two compounds.[27] It should be noted that attempts to prepare compounds with smaller amounts of the B ion than 0.25 were not successful.

Oxygen-deficient Oxides of the $A^{2+}(B^{3+}_{0.5}B^{4+}_{0.5})O_{2.75}$ and $A^{2+}(B^{2+}_{0.5}B^{5+}_{0.5})O_{2.75}$ Types

All compounds with the general formula $A^{2+}(B'_{0.5}B''_{0.5})O_{2.75}$ were found to have the ordered structure. Compounds containing tantalum as the B″ ions were first reported by Brixner.[24] Later uranium and molybdenum were also found to form these oxygen-deficient compounds.[17, 28]

A summary of the oxides with the perovskite structure and structural data are presented in Table 2.2.

TABLE 2.2. *Unit Cell Data for Perovskite-type Compounds* $A^{1+}B^{5+}O_3$

Compound	a (Å)	b (Å)	c (Å)	Remarks	References
$AgNbO_3$	7.888	15.660	7.888	$\beta = 90.57°$	29
$AgTaO_3$	3.931	3.914	3.931	$\beta = 90.35°$	29
$CsIO_3$	9.324 } or 4.66			cubic	30, 31
KIO_3	8.92 } or 4.410			cubic; $\alpha = 80°24'$ rhombohedral	4, 31, 32, 33, 34
$KNbO_3$	3.9714	5.6946	5.7203	orthorhombic	35
$KTaO_3$	3.9885			cubic	31, 36, 37
$NaNbO_3$	5.512	5.577	3.885	orthorhombic	35, 38
$NaTaO_3$	3.8851	5.4778	5.5239	orthorhombic	39
$RbIO_3$	4.52 } or 9.04			cubic	2, 31
$TlIO_3$	4.510			$\alpha = 89.34°$ rhombohedral	4, 40, 41

$A^{2+}B^{4+}O_3$

Compound	a (Å)	b (Å)	c (Å)	Remarks	References
$BaCeO_3$	4.397				4, 5, 35, 42
$BaFeO_3$	3.98		4.01	tetragonal	43, 44
$BaMoO_3$	4.0404				45
$BaPbO_3$	4.273				46
$BaPrO_3$	8.708 } or 4.354				31, 47
$BaPuO_3$	4.39				48
$BaSnO_3$	4.117				1, 4, 5
$BaThO_3$	4.480 8.985			monoclinic	4, 35, 42, 47
$BaTiO_3$	3.989		4.029	tetragonal	5
$BaUO_3$	4.387			pseudocubic	49
$BaZrO_3$	4.192			cubic	5
$CaCeO_3$	7.70			cubic	31, 50
$CaHfO_3$				orthorhombic	51
$CaMnO_3$	10.683	7.449	10.476	orthorhombic	13, 14, 52
$CaMoO_3$	7.80	7.77	7.80	$\beta = 91°23'$ monoclinic	53

TABLE 2.2 *(cont.)*

$A^{2+}B^{4+}O_3$ *(cont.)*

Compound	a (Å)	b (Å)	c (Å)	Remarks	References
$CaSnO_3$	5.518	7.884	5.664	orthorhombic	4, 54
$CaThO_3$	8.74			monoclinic, pseudocubic	31
$CaTiO_3$	5.381	7.645	5.443	orthorhombic	54
$CaUO_3$	5.78	8.29	5.97	orthorhombic	49
$CaVO_3$	5.326	7.547	5.352	orthorhombic	52
$CaZrO_3$	5.587	8.008	5.758	orthorhombic	55
$CdCeO_3$	7.65			orthorhombic or cubic	31, 50
$CdSnO_3$	5.547	5.577	7.867	orthorhombic	5
$CdThO_3$	8.74			pseudocubic	31
$CdTiO_3$	5.301	7.606	5.419	orthorhombic	5
$CdZrO_3$				orthorhombic	31
$EuTiO_3$	3.897				56
$MgCeO_3$	8.54				31
$PbCeO_3$	7.62			orthorhombic	31, 50
$PbHfO_3$				pseudotetragonal	57
$PbSnO_3$	7.86		8.13	tetragonal	31
$PbTiO_3$	3.896		4.136	tetragonal	58
$PbZrO_3$	9.28			pseudocubic, orthorhombic	1
$SrCeO_3$	5.986	8.531	6.125	orthorhombic	4, 5, 31
$SrCoO_3$	7.725				13
$SrFeO_3$	3.869				13
$SrHfO_3$	4.069 } or 8.138			orthorhombic	31
$SrMoO_3$	3.9751				45
$SrPbO_3$	5.864	5.949	8.336	orthorhombic	46
$SrRuO_3$				cubic	59
$SrSnO_3$	4.0334 } or 8.070			cubic	1, 4, 60
$SrThO_3$	8.84			pseudocubic	31
$SrTiO_3$	3.904			cubic	5
$SrUO_3$	6.01	8.60	6.17	orthorhombic	49
$SrZrO_3$	5.792 } or 8.218	8.189	5.818	orthorhombic cubic	4, 5, 31

$A^{3+}B^{3+}O_3$

Compound	a (Å)	b (Å)	c (Å)	Remarks	References
$BiAlO_3$	7.61		7.94	tetragonal	31

TABLE 2.2 *(cont.)*

$A^{3+}B^{3+}O_3$ *(cont.)*

Compound	a (Å)	b (Å)	c (Å)	Remarks	References
$BiCrO_3$	3.90	3.87	3.90	$\alpha = \gamma = 90°35'$ triclinic $\beta = 89°10'$	61
$BiMnO_3$	3.93	3.98	3.93	$\alpha = \gamma = 91°25'$ triclinic $\beta = 90°55'$	61
$CeAlO_3$	3.767		3.794	tetragonal	62, 63
$CeCrO_3$	3.866				63, 64, 65
$CeFeO_3$	3.900			pseudocubic, orthorhombic	64, 63
$CeGaO_3$	3.879			cubic, orthorhombic	64
$CeScO_3$				orthorhombic	64
$CeVO_3$	3.90				65, 66
$CrBiO_3$	7.77		8.08	tetragonal	31
$DyAlO_3$	5.21	5.31	7.40	orthorhombic	67
$DyFeO_3$	5.30	5.60	7.62	orthorhombic	67
$DyMnO_3$	3.70			cubic	68
$EuAlO_3$	5.271	5.292	7.458	$GdFeO_3$ structure	63, 69
$EuCrO_3$	3.803				
$EuFeO_3$	5.371	5.611	7.686	$GdFeO_3$ structure	63, 6, 63, 69
$FeBiO_3$	7.64 } or 3.965			$\alpha = 89°28'$ rhombohedral	70, 71, 72
$GdAlO_3$	5.247	5.304	7.447	$GdFeO_3$ structure	63, 67, 69
$GdCoO_3$	3.732	3.807	3.676	orthorhombic	73
$GdCrO_3$	5.312	5.514	7.611	$GdFeO_3$ structure	74
$GdFeO_3$	5.346	5.616	7.668	orthorhombic	6, 75
$GdMnO_3$	3.82				14
$GdScO_3$	5.487	5.756	7.925	$GdFeO_3$ structure	74
$GdVO_3$	5.345	5.623	7.638	$GdFeO_3$ structure	74
$LaAlO_3$	3.788			$\alpha = 90°4'$ rhombohedral	5

STRUCTURE OF PEROVSKITE-TYPE COMPOUNDS

TABLE 2.2 *(cont.)* $A^{3+}B^{3+}O_3$ *(cont.)*

Compound	a (Å)	b (Å)	c (Å)	Remarks	References
$LaCoO_3$	3.824 } or 7.651			$\alpha = 90°42'$ rhombohedral	13, 63, 65, 76, 77, 78
$LaCrO_3$	5.477	5.514	7.755	$GdFeO_3$ structure	74
$LaFeO_3$	5.556	5.565	7.862	$GdFeO_3$ structure	5, 6
$LaGaO_3$	5.496	5.524	7.787	$GdFeO_3$ structure	74
$LaInO_3$	5.723	8.207	5.914	orthorhombic	5, 64
$LaNiO_3$	7.676			$\alpha = 90°41'$ pseudocubic	78, 79
$LaRhO_3$	3.94				76, 80
$LaScO_3$	5.678	5.787	8.098	$GdFeO_3$ structure	64, 74
$LaTiO_3$	3.92				81, 82
$LaVO_3$	3.99 } or 7.842			cubic	65, 66, 83
$LaYO_3$				orthorhombic	84
$NdAlO_3$	3.752			rhombohedral	63, 67, 69
$NdCoO_3$	3.777				65, 73
$NdCrO_3$	5.412	5.494	7.695	$GdFeO_3$ structure	74
$NdFeO_3$	5.441	5.573	7.753	$GdFeO_3$ structure	6, 64
$NdGaO_3$	5.426	5.502	7.706	$GdFeO_3$ structure	74
$NdInO_3$	5.627	8.121	5.891	orthorhombic	5, 64
$NdMnO_3$	3.80				68
$NdScO_3$	5.574	5.771	7.998	$GdFeO_3$ structure	64, 74
$NdVO_3$	5.440	5.589	7.733	$GdFeO_3$ structure	74
$PrAlO_3$	3.757 } or 5.31			$\alpha = 60°20'$ rhombohedral	48, 63, 69
$PrCoO_3$	3.787			$\alpha = 90°13'$ rhombohedral	65, 73
$PrCrO_3$	5.444	5.484	7.710	$GdFeO_3$ structure	74

TABLE 2.2 *(cont.)*

$A^{3+}B^{3+}O_3$ *(cont.)*

Compound	a (Å)	b (Å)	c (Å)	Remarks	References
$PrFeO_3$	5.495	5.578	7.810	$GdFeO_3$ structure	74
$PrGaO_3$	5.465	5.495	7.729	$GdFeO_3$ structure	74
$PrMnO_3$	3.82				68
$PrScO_3$	5.615	5.776	8.027	$GdFeO_3$ structure	74
$PrVO_3$	5.477	5.545	7.759	$GdFeO_3$ structure	74
$PuAlO_3$	5.33			$\alpha = 56°4'$ rhombohedral	48
$PuCrO_3$	5.46	5.51	7.76	$GdFeO_3$ structure	48
$PuMnO_3$	3.86			pseudocubic	48
$PuVO_3$	5.48	5.61	7.78	$GdFeO_3$ structure	48
$SmAlO_3$	5.285	5.290	7.473	$GdFeO_3$ structure	64, 63, 69, 74
$SmCoO_3$	3.747	3.803	3.728	orthorhombic	65, 73
$SmCrO_3$	5.372	5.502	7.650	$GdFeO_3$ structure	74
$SmFeO_3$	5.394	5.592	7.711	$GdFeO_3$ structure	6, 64
$SmInO_3$	5.589	8.082	5.886	orthorhombic	5, 65
$SmVO_3$	3.89				
$YAlO_3$	5.179	5.329	7.370	$GdFeO_3$ structure	6
$YCrO_3$	5.247	5.518	7.540	$GdFeO_3$ structure	6, 7, 64, 74
$YFeO_3$	5.302	5.589	7.622	$GdFeO_3$ structure	65
$YScO_3$	5.431	5.712	7.894	$GdFeO_3$ structure	74

A_xBO_3 and ABO_{3-x}

Compound	a (Å)	b (Å)	c (Å)	Remarks	References
$Ce_{0.33}NbO_3$	3.89	3.91	7.86	orthorhombic	10
$Ce_{0.33}TaO_3$	3.90	3.91	7.86	orthorhombic	10

STRUCTURE OF PEROVSKITE-TYPE COMPOUNDS

TABLE 2.2 *(cont.)*

A_xBO_3 and ABO_{3-x} *(cont.)*

Compound	a (Å)	b (Å)	c (Å)	Remarks	References
$Dy_{0.33}TaO_3$	3.83	3.83	7.75	$\gamma = 90.8°$ monoclinic	10
$Gd_{0.33}TaO_3$	3.87	3.89	7.73	orthorhombic	10
$La_{0.33}NbO_3$	3.91		7.90	tetragonal	10
$La_{0.33}TaO_3$	3.92		7.88	tetragonal	10
$Nd_{0.33}NbO_3$	3.90	3.91	7.76	orthorhombic	10
$Nd_{0.33}TaO_3$	3.91		7.77	tetragonal	10
$Pr_{0.33}NbO_3$	3.91	3.92	7.77	orthorhombic	10
$Pr_{0.33}TaO_3$	3.91	3.92	7.78	orthorhombic	10
$Sm_{0.33}TaO_3$	3.89		7.75	tetragonal	10
$Y_{0.33}TaO_3$	3.82	3.83	7.74	$\gamma = 90.9°$ monoclinic	10
$Yb_{0.33}TaO_3$	3.79	3.80	7.70	$\gamma = 91.6°$ monoclinic	10
$Ca_{0.5}TaO_3$	11.068	7.505	5.378	orthorhombic	85
Li_xWO_3	($x=1$) 3.72			cubic $x = 0.35$–0.57	86
Na_xWO_3	($x=1$) 3.8622			cubic $x = 0.7$–1.0	8, 50, 87, 88, 89, 80, 91, 92
$Sr_{0.5+x}Nb^{4+}_{2x}Nb^{5+}_{1-2x}O_3$	($x=0.2$) 3.981 ($x=0.45$) 4.016			cubic $x = 0.7$–0.9	9 9
$CaMnO_{3-x}$					13, 14
$SrCoO_{3-x}$					13
$SrFeO_{3-x}$					11, 12, 13
$SrTiO_{3-x}$					83
$SrVO_{3-x}$					83

$A(B'_{0.67}B''_{0.33})O_3$					
$Ba(Al_{0.67}W_{0.33})O_3$					93
$Ba(Dy_{0.67}W_{0.33})O_3$	8.386			$(NH_4)_3FeF_6$ structure	93

TABLE 2.2 *(cont.)* $A(B'_{0.67}B''_{0.33})O_3$ *(cont.)*

Compound	a (Å)	b (Å)	c (Å)	Remarks	References
$Ba(Er_{0.67}W_{0.33})O_3$	8.386			$(NH_4)_3FeF_6$ structure	93
$Ba(Eu_{0.67}W_{0.33})O_3$	8.605			$(NH_4)_3FeF_6$ structure	93
$Ba(Fe_{0.67}U_{0.33})O_3$	8.232			$(NH_4)_3FeF_6$ structure	17
$Ba(Gd_{0.67}W_{0.33})O_3$	8.411			$(NH_4)_3FeF_6$ structure	93, 94
$Ba(In_{0.67}U_{0.33})O_3$	8.512			$(NH_4)_3FeF_6$ structure	17
$Ba(In_{0.67}W_{0.33})O_3$	8.321			ordered structure	93
$Ba(La_{0.67}W_{0.33})O_3$	8.58			$(NH_4)_3FeF_6$ structure	94
$Ba(Lu_{0.67}W_{0.33})O_3$					93
$Ba(Nd_{0.67}W_{0.33})O_3$	8.513			$(NH_4)_3FeF_6$ structure	93
$Ba(Sc_{0.67}U_{0.33})O_3$	8.49			$(NH_4)_3FeF_6$ structure	17
$Ba(Sc_{0.67}W_{0.33})O_3$	8.24			$(NH_4)_3FeF_6$ structure	15, 93
$Ba(Y_{0.67}U_{0.33})O_3$	8.70			$(NH_4)_3FeF_6$ structure	17
$Ba(Y_{0.67}W_{0.33})O_3$	8.374			$(NH_4)_3FeF_6$ structure	93
$Ba(Yb_{0.67}W_{0.33})O_3$					93
$Pb(Fe_{0.67}W_{0.33})O_3$					95, 96
$Sr(Cr_{0.67}Re_{0.33})O_3$	8.01			$(NH_4)_3FeF_6$ structure	18
$Sr(Cr_{0.67}U_{0.33})O_3$	8.00			$(NH_4)_3FeF_6$ structure	17
$Sr(Fe_{0.67}Re_{0.33})O_3$	7.89			$(NH_4)_3FeF_6$ structure	18
$Sr(Fe_{0.67}W_{0.33})O_3$	3.945		3.951	tetragonal	47, 94
$Sr(In_{0.67}Re_{0.33})O_3$	8.297			$(NH_4)_3FeF_6$ structure	18
$La(Co_{0.67}Nb_{0.33})O_3$	5.58	5.58	7.89	orthorhombic	94
$La(Co_{0.67}Sb_{0.33})O_3$	5.57	5.57	7.89	orthorhombic	94

TABLE 2.2 (cont.) $A^{2+}(B^{2+}_{0.33}B^{5+}_{0.67})O_3$ (cont.)

Compound	a (Å)	b (Å)	c (Å)	Remarks	References
$Ba(Ca_{0.33}Nb_{0.67})O_3$	5.92		7.25	hex. ordered $Ba(Sr_{0.33}Ta_{0.67})O_3$ structure	23
$Ba(Ca_{0.33}Ta_{0.67})O_3$	5.895		7.284	hex. ordered $Ba(Sr_{0.33}Ta_{0.67})O_3$ structure	21, 97, 98
$Ba(Cd_{0.33}Nb_{0.67})O_3$	4.168				23
$Ba(Cd_{0.33}Ta_{0.67})O_3$	4.167				22
$Ba(Co_{0.33}Nb_{0.33})O_3$	4.09				20
$Ba(Co_{0.33}Ta_{0.67})O_3$	5.776		7.082	hex. ordered $Ba(Sr_{0.33}Ta_{0.67})O_3$ structure	19, 22
$Ba(Cu_{0.33}Nb_{0.67})O_3$	8.04		8.40	tetragonal	94
$Ba(Fe_{0.33}Nb_{0.67})O_3$	4.085				23
$Ba(Fe_{0.33}Ta_{0.67})O_3$	4.10				20
$Ba(Mg_{0.33}Nb_{0.67})O_3$	5.77		7.08	hex. ordered $Ba(Sr_{0.33}Ta_{0.67})O_3$ structure	20, 23, 94
$Ba(Mg_{0.33}Ta_{0.67})O_3$	5.782		7.067	hex. ordered $Ba(Sr_{0.33}Ta_{0.67})O_3$ structure	22, 98
$Ba(Mn_{0.33}Nb_{0.67})O_3$					93
$Ba(Mn_{0.33}Ta_{0.67})O_3$	5.819		7.127	hex. ordered $Ba(Sr_{0.33}Ta_{0.67})O_3$ structure	22
$Ba(Ni_{0.33}Nb_{0.67})O_3$	4.074				19, 23, 96
$Ba(Ni_{0.33}Ta_{0.67})O_3$	5.758		7.052	hex. ordered $Ba(Sr_{0.33}Ta_{0.67})O_3$ structure	19, 22, 98
$Ba(Pb_{0.33}Nb_{0.67})O_3$	4.26				23
$Ba(Pb_{0.33}Ta_{0.67})O_3$	4.25				22
$Ba(Sr_{0.33}Ta_{0.67})O_3$	5.95		7.47	hex. ordered $Ba(Sr_{0.33}Ta_{0.67})O_3$ structure	20, 21
$Ba(Zn_{0.33}Nb_{0.67})O_3$	4.094				20, 23, 96
$Ba(Zn_{0.33}Ta_{0.67})O_3$	5.782		7.097	hex. ordered $Ba(Sr_{0.33}Ta_{0.67})O_3$ structure	20, 22, 98
$Ca(Ni_{0.33}Nb_{0.67})O_3$	3.88				96
$Ca(Ni_{0.33}Ta_{0.67})O_3$	3.93				19
$Pb(Co_{0.33}Nb_{0.67})O_3$	4.04				96, 99
$Pb(Co_{0.33}Ta_{0.67})O_3$	4.01				99

TABLE 2.2 (cont.) $A^{2+}(B^{2+}_{0.33}B^{5+}_{0.67})O_3$ (cont.)

Compound	a (Å)	b (Å)	c (Å)	Remarks	References
$Pb(Mg_{0.33}Nb_{0.67})O_3$	4.041				96, 100
$Pb(Mg_{0.33}Ta_{0.67})O_3$	4.02				96, 99
$Pb(Mn_{0.33}Nb_{0.67})O_3$					96
$Pb(Ni_{0.33}Nb_{0.67})O_3$	4.025				96, 100
$Pb(Ni_{0.33}Ta_{0.67})O_3$	4.01				96, 99
$Pb(Zn_{0.33}Nb_{0.67})O_3$	4.04				99
$Sr(Ca_{0.33}Nb_{0.67})O_3$	5.76		7.16	hex. ordered $Ba(Sr_{0.33}Ta_{0.67})O_3$ structure	23
$Sr(Ca_{0.33}Sb_{0.67})O_3$	8.17			$(NH_4)_3FeF_6$ structure	94
$Sr(Ca_{0.33}Ta_{0.67})O_3$	5.764		7.096	hex. ordered $Ba(Sr_{0.33}Ta_{0.67})O_3$ structure	22
$Sr(Cd_{0.33}Nb_{0.67})O_3$	4.089				23
$Sr(Co_{0.33}Nb_{0.67})O_3$	8.01			$(NH_4)_3FeF_6$ structure	94
$Sr(Co_{0.33}Sb_{0.67})O_3$	7.99			$(NH_4)_3FeF_6$ structure	94
$Sr(Co_{0.33}Ta_{0.67})O_3$	5.630		6.937	hex. ordered $Ba(Sr_{0.33}Ta_{0.67})O_3$ structure	20, 22
$Sr(Cu_{0.33}Sb_{0.67})O_3$	7.84		8.19	tetragonal	94
$Sr(Fe_{0.33}Nb_{0.67})O_3$	3.997		4.018	tetragonal	23
$Sr(Mg_{0.33}Nb_{0.67})O_3$	5.66		6.98	hex. ordered $Ba(Sr_{0.33}Ta_{0.67})O_3$ structure	23
$Sr(Mg_{0.33}Sb_{0.67})O_3$	7.96			$(NH_4)_3FeF_6$ structure	94
$Sr(Mg_{0.33}Ta_{0.67})O_3$	5.652		6.951	hex. ordered $Ba(Sr_{0.33}Ta_{0.67})O_3$ structure	19, 22
$Sr(Mn_{0.33}Nb_{0.67})O_3$					93
$Sr(Mn_{0.33}Ta_{0.67})O_3$					93
$Sr(Ni_{0.33}Nb_{0.67})O_3$	5.64		6.90	hex. ordered $Ba(Sr_{0.33}Ta_{0.67})O_3$ structure	23, 96
$Sr(Ni_{0.33}Ta_{0.67})O_3$	5.607		6.923	hex. ordered $Ba(Sr_{0.33}Ta_{0.67})O_3$ structure	20, 22
$Sr(Pb_{0.33}Nb_{0.67})O_3$					93
$Sr(Pb_{0.33}Ta_{0.67})O_3$					93

STRUCTURE OF PEROVSKITE-TYPE COMPOUNDS

TABLE 2.2 (cont.) $A^{2+}(B^{2+}_{0.33}B^{5+}_{0.67})O_3$ (cont.)

Compound	a (Å)	b (Å)	c (Å)	Remarks	References
$Sr(Zn_{0.33}Nb_{0.67})O_3$	5.66		6.95	hex. ordered $Ba(Sr_{0.33}Ta_{0.67})O_3$ structure	20, 23
$Sr(Zn_{0.33}Ta_{0.67})O_3$	5.664		6.951	hex. ordered $Ba(Sr_{0.33}Ta_{0.67})O_3$ structure	20, 22

$A^{2+}(B^{3+}_{0.5}B^{5+}_{0.5})O_3$

Compound	a (Å)	b (Å)	c (Å)	Remarks	References
$Ba(Bi_{0.5}Nb_{0.5})O_3$	8.630			$(NH_4)_3FeF_6$ structure	101
$Ba(Bi_{0.5}Ta_{0.5})O_3$	8.568			$(NH_4)_3FeF_6$ structure	101
$Ba(Ce_{0.5}Nb_{0.5})O_3$	4.293				102
$Ba(Ce_{0.5}Pa_{0.5})O_3$	8.800			$(NH_4)_3FeF_6$ structure	103
$Ba(Co_{0.5}Nb_{0.5})O_3$	4.06			$(NH_4)_3FeF_6$ structure	94
$Ba(Co_{0.5}Re_{0.5})O_3$	8.086			$(NH_4)_3FeF_6$ structure	18
$Ba(Cr_{0.5}W_{0.5})O_3$					104
$Ba(Cu_{0.5}W_{0.5})O_3$	7.88		8.61	tetragonal	94
$Ba(Dy_{0.5}Nb_{0.5})O_3$	8.437			$(NH_4)_3FeF_6$ structure	25, 102
$Ba(Dy_{0.5}Pa_{0.5})O_3$	8.740			$(NH_4)_3FeF_6$ structure	103
$Ba(Dy_{0.5}Ta_{0.5})O_3$	8.545			$(NH_4)_3FeF_6$ structure	105
$Ba(Er_{0.5}Nb_{0.5})O_3$	8.427			$(NH_4)_3FeF_6$ structure	25, 102
$Ba(Er_{0.5}Pa_{0.5})O_3$	8.716			$(NH_4)_3FeF_6$ structure	103
$Ba(Er_{0.5}Re_{0.5})O_3$	8.354			$(NH_4)_3FeF_6$ structure	18
$Ba(Er_{0.5}Ta_{0.5})O_3$	8.423			$(NH_4)_3FeF_6$ structure	105
$Ba(Er_{0.5}U_{0.5})O_3$	8.67			$(NH_4)_3FeF_6$ structure	104
$Ba(Eu_{0.5}Nb_{0.5})O_3$	8.507			$(NH_4)_3FeF_6$ structure	25, 102
$Ba(Eu_{0.5}Pa_{0.5})O_3$	8.783			$(NH_4)_3FeF_6$ structure	103

TABLE 2.2 (cont.) $A^{2+}(B^{3+}_{0.5}B^{5+}_{0.5})O_3$ (cont.)

Compound	a (Å)	b (Å)	c (Å)	Remarks	References
$Ba(Eu_{0.5}Ta_{0.5})O_3$	8.506				105
$Ba(Fe_{0.5}Mo_{0.5})O_3$	8.08			$(NH_4)_3FeF_6$ structure	28
$Ba(Fe_{0.5}Nb_{0.5})O_3$	4.06				20, 25, 106
$Ba(Fe_{0.5}Re_{0.5})O_3$	8.05			$(NH_4)_3FeF_6$ structure	18
$Ba(Fe_{0.5}Ta_{0.5})O_3$	4.056				20, 103
$Ba(Gd_{0.5}Nb_{0.5})O_3$	8.496			$(NH_4)_3FeF_6$ structure	25, 102
$Ba(Gd_{0.5}Pa_{0.5})O_3$	8.774			$(NH_4)_3FeF_6$ structure	103
$Ba(Gd_{0.5}Re_{0.5})O_3$	8.431			$(NH_4)_3FeF_6$ structure	18
$Ba(Gd_{0.5}Sb_{0.5})O_3$	8.44			$(NH_4)_3FeF_6$ structure	94
$Ba(Gd_{0.5}Ta_{0.5})O_3$	8.487		8.513	tetragonal	105
$Ba(Ho_{0.5}Nb_{0.5})O_3$	8.434			$(NH_4)_3FeF_6$ structure	25, 102
$Ba(Ho_{0.5}Pa_{0.5})O_3$	8.730			$(NH_4)_3FeF_6$ structure	103
$Ba(Ho_{0.5}Ta_{0.5})O_3$	8.442			$(NH_4)_3FeF_6$ structure	105
$Ba(In_{0.5}Nb_{0.5})O_3$	8.279			$(NH_4)_3FeF_6$ structure	25, 107
$Ba(In_{0.5}Os_{0.5})O_3$	8.224			$(NH_4)_3FeF_6$ structure	18
$Ba(In_{0.5}Pa_{0.5})O_3$	8.596			$(NH_4)_3FeF_6$ structure	103
$Ba(In_{0.5}Re_{0.5})O_3$	8.258			$(NH_4)_3FeF_6$ structure	18
$Ba(In_{0.5}Sb_{0.5})O_3$	8.269			$(NH_4)_3FeF_6$ structure	94, 108
$Ba(In_{0.5}Ta_{0.5})O_3$	8.280			$(NH_4)_3FeF_6$ structure	93
$Ba(In_{0.5}U_{0.5})O_3$	8.52			$(NH_4)_3FeF_6$ structure	104
$Ba(La_{0.5}Nb_{0.5})O_3$	8.607		8.690	tetragonal	25, 102, 106
$Ba(La_{0.5}Pa_{0.5})O_3$	8.885			$(NH_4)_3FeF_6$ structure	103

TABLE 2.2 (cont.) $A^{2+}(B^{3+}_{0.5}B^{5+}_{0.5})O_3$ (cont.)

Compound	a (Å)	b (Å)	c (Å)	Remarks	References
$Ba(La_{0.5}Re_{0.5})O_3$	8.58			$(NH_4)_3FeF_6$ structure	18
$Ba(La_{0.5}Ta_{0.5})O_3$	8.611	8.639	8.764	$(NH_4)_3FeF_6$ structure	24, 93, 106
$Ba(Lu_{0.5}Nb_{0.5})O_3$	8.364			$(NH_4)_3FeF_6$ structure	25, 102
$Ba(Lu_{0.5}Pa_{0.5})O_3$	8.666			$(NH_4)_3FeF_6$ structure	103
$Ba(Lu_{0.5}Ta_{0.5})O_3$	8.372			$(NH_4)_3FeF_6$ structure	105
$Ba(Mn_{0.5}Nb_{0.5})O_3$	4.083				106
$Ba(Mn_{0.5}Re_{0.5})O_3$	8.18			$(NH_4)_3FeF_6$ structure	18
$Ba(Mn_{0.5}Ta_{0.5})O_3$	4.076				106
$Ba(Nd_{0.5}Nb_{0.5})O_3$	8.540			$(NH_4)_3FeF_6$ structure	25, 102, 106
$Ba(Nd_{0.5}Pa_{0.5})O_3$	8.840			$(NH_4)_3FeF_6$ structure	103
$Ba(Nd_{0.5}Re_{0.5})O_3$	8.51			$(NH_4)_3FeF_6$ structure	18
$Ba(Nd_{0.5}Ta_{0.5})O_3$	8.556				105, 106
$Ba(Ni_{0.5}Nb_{0.5})O_3$	4.1				94
$Ba(Pr_{0.5}Nb_{0.5})O_3$	4.27				102, 106
$Ba(Pr_{0.5}Pa_{0.5})O_3$	8.862			$(NH_4)_3FeF_6$ structure	103
$Ba(Pr_{0.5}Ta_{0.5})O_3$	4.27			$(NH_4)_3FeF_6$ structure	108
$Ba(Rh_{0.5}Nb_{0.5})O_3$	8.17			structure	94
$Ba(Rh_{0.5}U_{0.5})O_3$				hexagonal $BaTiO_3$	104
$Ba(Sc_{0.5}Nb_{0.5})O_3$	4.121				96, 102
$Ba(Sc_{0.5}Os_{0.5})O_3$	8.152			$(NH_4)_3FeF_6$ structure	18
$Ba(Sc_{0.5}Pa_{0.5})O_3$	8.549			$(NH_4)_3FeF_6$ structure	103
$Ba(Sc_{0.5}Re_{0.5})O_3$	8.163			$(NH_4)_3FeF_6$ structure	18
$Ba(Sc_{0.5}Sb_{0.5})O_3$	8.197			$(NH_4)_3FeF_6$ structure	108
$Ba(Sc_{0.5}Ta_{0.5})O_3$	8.222			$(NH_4)_3FeF_6$ structure	93, 96

TABLE 2.2 (cont.) $A^{2+}(B^{3+}_{0.5}B^{5+}_{0.5})O_3$ (cont.)

Compound	a (Å)	b (Å)	c (Å)	Remarks	References
$Ba(Sc_{0.5}U_{0.5})O_3$	8.49			$(NH_4)_3FeF_6$ structure	104
$Ba(Sm_{0.5}Nb_{0.5})O_3$	8.518			$(NH_4)_3FeF_6$ structure	25, 102, 106
$Ba(Sm_{0.5}Pa_{0.5})O_3$	8.792			$(NH_4)_3FeF_6$ structure	103
$Ba(Sm_{0.5}Ta_{0.5})O_3$	8.519				105, 106
$Ba(Tb_{0.5}Nb_{0.5})O_3$	4.229				102
$Ba(Tb_{0.5}Pa_{0.5})O_3$	8.753			$(NH_4)_3FeF_6$ structure	103
$Ba(Tl_{0.5}Ta_{0.5})O_3$	8.42			$(NH_4)_3FeF_6$ structure	108
$Ba(Tm_{0.5}Nb_{0.5})O_3$	8.408			$(NH_4)_3FeF_6$ structure	25, 102
$Ba(Tm_{0.5}Pa_{0.5})O_3$	8.692			$(NH_4)_3FeF_6$ structure	103
$Ba(Tm_{0.5}Ta_{0.5})O_3$	8.406			$(NH_4)_3FeF_6$ structure	105
$Ba(Y_{0.5}Nb_{0.5})O_3$	4.200				102, 106
$Ba(Y_{0.5}Pa_{0.5})O_3$	8.718			$(NH_4)_3FeF_6$ structure	103
$Ba(Y_{0.5}Re_{0.5})O_3$	8.372			$(NH_4)_3FeF_6$ structure	18
$Ba(Y_{0.5}Ta_{0.5})O_3$	8.433			$(NH_4)_3FeF_6$ structure	105, 106
$Ba(Y_{0.5}U_{0.5})O_3$	8.69			$(NH_4)_3FeF_6$ structure	104
$Ba(Yb_{0.5}Nb_{0.5})O_3$	8.374			$(NH_4)_3FeF_6$ structure	25, 96, 102
$Ba(Yb_{0.5}Pa_{0.5})O_3$	8.678			$(NH_4)_3FeF_6$ structure	103
$Ba(Yb_{0.5}Ta_{0.5})O_3$	8.390			$(NH_4)_3FeF_6$ structure	96, 105
$Ca(Al_{0.5}Nb_{0.5})O_3$	3.81	3.80	3.81	$\beta = 90°15'$ monoclinic	109
$Ca(Al_{0.5}Ta_{0.5})O_3$	3.81	3.80	3.81	$\beta = 90°17'$ monoclinic	109
$Ca(Co_{0.5}W_{0.5})O_3$	5.60	5.43	7.73	orthorhombic	94
$Ca(Cr_{0.5}Mo_{0.5})O_3$	5.49	7.70	5.36	orthorhombic	28

TABLE 2.2 (cont.) $A^{2+}(B^{3+}_{0.5}B^{5+}_{0.5})O_3$ (cont.)

Compound	a(Å)	b(Å)	c(Å)	Remarks	References
$Ca(Cr_{0.5}Nb_{0.5})O_3$	3.85	3.85	3.85	$\beta = 90°47'$ monoclinic	109
$Ca(Cr_{0.5}Os_{0.5})O_3$	5.38	7.66	5.47	orthorhombic	18
$Ca(Cr_{0.5}Re_{0.5})O_3$	5.38	7.67	5.47	orthorhombic	18
$Ca(Cr_{0.5}Ta_{0.5})O_3$	3.85	3.85	3.85	$\beta = 90°45'$ monoclinic	109
$Ca(Cr_{0.5}W_{0.5})O_3$	5.47	7.70	5.35	orthorhombic	28
$Ca(Dy_{0.5}Nb_{0.5})O_3$	4.03	4.03	4.03	$\beta = 92°25'$ monoclinic	109
$Ca(Dy_{0.5}Ta_{0.5})O_3$	4.03	4.03	4.03	$\beta = 92°24'$ monoclinic	109
$Ca(Er_{0.5}Nb_{0.5})O_3$	4.02	4.01	4.02	$\beta = 92°11'$ monoclinic	109
$Ca(Er_{0.5}Ta_{0.5})O_3$	4.02	4.01	4.02	$\beta = 92°10'$ monoclinic	109
$Ca(Fe_{0.5}Mo_{0.5})O_3$	5.53	7.73	5.42	orthorhombic	28
$Ca(Fe_{0.5}Nb_{0.5})O_3$	3.89	3.88	3.89	$\beta = 91°2'$ monoclinic	109
$Ca(Fe_{0.5}Sb_{0.5})O_3$	5.54	5.47	7.74	orthorhombic	94
$Ca(Fe_{0.5}Ta_{0.5})O_3$	3.89	3.88	3.89	$\beta = 91°7'$ monoclinic	109
$Ca(Gd_{0.5}Nb_{0.5})O_3$	4.03	4.04	4.03	$\beta = 92°42'$ monoclinic	109
$Ca(Gd_{0.5}Ta_{0.5})O_3$	4.03	4.04	4.05	$\beta = 92°41'$ monoclinic	109
$Ca(Ho_{0.5}Nb_{0.5})O_3$	4.02	4.02	4.02	$\beta = 92°19'$ monoclinic	109
$Ca(Ho_{0.5}Ta_{0.5})O_3$	4.03	4.02	4.03	$\beta = 92°16'$ monoclinic	109
$Ca(In_{0.5}Nb_{0.5})O_3$	3.97	3.95	3.97	$\beta = 91°53'$ monoclinic	109
$Ca(In_{0.5}Ta_{0.5})O_3$	3.97	3.96	3.97	$\beta = 91°51'$ monoclinic	109
$Ca(La_{0.5}Nb_{0.5})O_3$	4.07	4.07	4.07	$\beta = 92°8'$ monoclinic	109
$Ca(La_{0.5}Ta_{0.5})O_3$	4.07	4.07	4.07	$\beta = 92°9'$ monoclinic	109
$Ca(Mn_{0.5}Ta_{0.5})O_3$	3.90	3.87	3.90	$\beta = 91°9'$ monoclinic	109
$Ca(Nd_{0.5}Nb_{0.5})O_3$	4.05	4.05	4.05	$\beta = 92°28'$ monoclinic	109

TABLE 2.2 (cont.) $A^{2+}(B^{3+}_{0.5}B^{5+}_{0.5})O_3$ (cont.)

Compound	a (Å)	b (Å)	c (Å)	Remarks	References
$Ca(Nd_{0.5}Ta_{0.5})O_3$	4.05	4.05	4.05	$\beta = 92°25'$ monoclinic	109
$Ca(Ni_{0.5}W_{0.5})O_3$	5.55	5.40	7.70	orthorhombic	94
$Ca(Pr_{0.5}Nb_{0.5})O_3$	4.06	4.05	4.06	$\beta = 92°25'$ monoclinic	109
$Ca(Pr_{0.5}Ta_{0.5})O_3$	4.06	4.05	4.06	$\beta = 92°22'$ monoclinic	109
$Ca(Sc_{0.5}Re_{0.5})O_3$	5.49	7.86	5.63	orthorhombic	18
$Ca(Sm_{0.5}Nb_{0.5})O_3$	4.04	4.04	4.04	$\beta = 92°42'$ monoclinic	109
$Ca(Sm_{0.5}Ta_{0.5})O_3$	4.05	4.04	4.05	$\beta = 92°28'$ monoclinic	109
$Ca(Tb_{0.5}Nb_{0.5})O_3$	4.03	4.03	4.03	$\beta = 92°35'$ monoclinic	109
$Ca(Tb_{0.5}Ta_{0.5})O_3$	4.03	4.03	4.03	$\beta = 92°36'$ monoclinic	109
$Ca(Y_{0.5}Nb_{0.5})O_3$	4.03	4.02	4.03	$\beta = 92°23'$ monoclinic	109
$Ca(Y_{0.5}Ta_{0.5})O_3$	4.03	4.02	4.03	$\beta = 92°23'$ monoclinic	109
$Ca(Yb_{0.5}Nb_{0.5})O_3$	4.01	4.00	4.01	$\beta = 92°0'$ monoclinic	109
$Ca(Yb_{0.5}Ta_{0.5})O_3$	4.01	4.00	4.01	$\beta = 92°3'$ monoclinic	109
$Pb(Fe_{0.5}Nb_{0.5})O_3$	4.017				26, 110
$Pb(Fe_{0.5}Ta_{0.5})O_3$	4.011				26
$Pb(In_{0.5}Nb_{0.5})O_3$	4.11				111
$Pb(Ho_{0.5}Nb_{0.5})O_3$	4.160		4.106	monoclinic	111
$Pb(Lu_{0.5}Nb_{0.5})O_3$	4.152		4.098	monoclinic	111
$Pb(Lu_{0.5}Ta_{0.5})O_3$	4.153		4.107	monoclinic	111
$Pb(Sc_{0.5}Nb_{0.5})O_3$	4.078		4.083	tetragonal	26, 112
$Pb(Sc_{0.5}Ta_{0.5})O_3$	4.072				26, 112
$Pb(Yb_{0.5}Nb_{0.5})O_3$	4.15				96, 111, 112
$Pb(Yb_{0.5}Ta_{0.5})O_3$	4.13				96, 111
$Sr(Co_{0.5}Nb_{0.5})O_3$	3.93				94
$Sr(Co_{0.5}Sb_{0.5})O_3$	7.88			$(NH_4)_3FeF_6$ structure	94
$Sr(Cr_{0.5}Mo_{0.5})O_3$	7.82			$(NH_4)_3FeF_6$ structure	28
$Sr(Cr_{0.5}Nb_{0.5})O_3$	3.9421				94, 107

TABLE 2.2 *(cont.)* $A^{2+}(B^{3+}_{0.5}B^{5+}_{0.5})O_3$ *(cont.)*

Compound	a (Å)	b (Å)	c (Å)	Remarks	References
$Sr(Cr_{0.5}Os_{0.5})O_3$	7.84			$(NH_4)_3FeF_6$ structure	18
$Sr(Cr_{0.5}Re_{0.5})O_3$	7.82			$(NH_4)_3FeF_6$ structure	18
$Sr(Cr_{0.5}Sb_{0.5})O_3$	7.862			$(NH_4)_3FeF_6$ structure	94, 108
$Sr(Cr_{0.5}Ta_{0.5})O_3$	3.94				19
$Sr(Cr_{0.5}W_{0.5})O_3$	7.82			$(NH_4)_3FeF_6$ structure	28
$Sr(Dy_{0.5}Ta_{0.5})O_3$					93
$Sr(Er_{0.5}Ta_{0.5})O_3$					93
$Sr(Eu_{0.5}Ta_{0.5})O_3$					93
$Sr(Fe_{0.5}Mo_{0.5})O_3$	7.89			$(NH_4)_3FeF_6$ structure	28
$Sr(Fe_{0.5}Nb_{0.5})O_3$	3.97				20
$Sr(Fe_{0.5}Sb_{0.5})O_3$	7.916			$(NH_4)_3FeF_6$ structure	94, 108
$Sr(Fe_{0.5}Ta_{0.5})O_3$	3.96		3.981	tetragonal	113
$Sr(Ga_{0.5}Nb_{0.5})O_3$	3.946				19
$Sr(Ga_{0.5}Os_{0.5})O_3$	7.82			$(NH_4)_3FeF_6$ structure	18
$Sr(Ga_{0.5}Re_{0.5})O_3$	7.843			$(NH_4)_3FeF_6$ structure	18
$Sr(Ga_{0.5}Sb_{0.5})O_3$	7.84		7.91	tetragonal	108
$Sr(Gd_{0.5}Ta_{0.5})O_3$					93
$Sr(Ho_{0.5}Ta_{0.5})O_3$					93
$Sr(In_{0.5}Nb_{0.5})O_3$	4.0569				107
$Sr(In_{0.5}Os_{0.5})O_3$	8.06			$(NH_4)_3FeF_6$ structure	18
$Sr(In_{0.5}Re_{0.5})O_3$	8.071			$(NH_4)_3FeF_6$ structure	18
$Sr(In_{0.5}U_{0.5})O_3$	8.33			$(NH_4)_3FeF_6$ structure	104
$Sr(La_{0.5}Ta_{0.5})O_3$	8.27			$(NH_4)_3FeF_6$ structure	24
$Sr(Lu_{0.5}Ta_{0.5})O_3$					93
$Sr(Mn_{0.5}Mo_{0.5})O_3$	7.98			$(NH_4)_3FeF_6$ structure	28
$Sr(Mn_{0.5}Sb_{0.5})O_3$					114
$Sr(Nd_{0.5}Ta_{0.5})O_3$					93
$Sr(Ni_{0.5}Sb_{0.5})O_3$				tetragonal	94

TABLE 2.2 (cont.)

$A^{2+}(B^{3+}_{0.5}B^{5+}_{0.5})O_3$ (cont.)

Compound	a (Å)	b (Å)	c (Å)	Remarks	References
$Sr(Rh_{0.5}Sb_{0.5})O_3$	5.77	5.55	7.99	orthorhombic	94
$Sr(Sc_{0.5}Os_{0.5})O_3$	8.02			$(NH_4)_3FeF_6$ structure	18
$Sr(Sc_{0.5}Re_{0.5})O_3$	8.02			$(NH_4)_3FeF_6$ structure	18
$Sr(Sm_{0.5}Ta_{0.5})O_3$					93
$Sr(Tm_{0.5}Ta_{0.5})O_3$					93
$Sr(Yb_{0.5}Ta_{0.5})O_3$					93

$A^{2+}(B^{2+}_{0.5}B^{6+}_{0.5})O_3$

Compound	a (Å)	b (Å)	c (Å)	Remarks	References
$Ba(Ba_{0.5}Os_{0.5})O_3$	8.66		8.34	tetragonal	18
$Ba(Ba_{0.5}Re_{0.5})O_3$	8.65		8.33	tetragonal	18
$Ba(Ba_{0.5}U_{0.5})O_3$	8.89			$(NH_4)_3FeF_6$ structure	17
$Ba(Ba_{0.5}W_{0.5})O_3$	8.6			$(NH_4)_3FeF_6$ structure	16
$Ba(Ca_{0.5}Mo_{0.5})O_3$	8.355			$(NH_4)_3FeF_6$ structure	16
$Ba(Ca_{0.5}Os_{0.5})O_3$	8.362			$(NH_4)_3FeF_6$ structure	18
$Ba(Ca_{0.5}Re_{0.5})O_3$	8.356			$(NH_4)_3FeF_6$ structure	18, 115
$Ba(Ca_{0.5}Te_{0.5})O_3$	8.393			$(NH_4)_3FeF_6$ structure	108
$Ba(Ca_{0.5}U_{0.5})O_3$	8.67			$(NH_4)_3FeF_6$ structure	17
$Ba(Ca_{0.5}W_{0.5})O_3$	8.39			$(NH_4)_3FeF_6$ structure	15, 16
$Ba(Cd_{0.5}Os_{0.5})O_3$	8.325			$(NH_4)_3FeF_6$ structure	18
$Ba(Cd_{0.5}Re_{0.5})O_3$	8.322			$(NH_4)_3FeF_6$ structure	18, 115
$Ba(Cd_{0.5}U_{0.5})O_3$	6.13	8.64	6.07	orthorhombic	17
$Ba(Co_{0.5}Mo_{0.5})O_3$	4.0429				107
$Ba(Co_{0.5}Re_{0.5})O_3$	8.086			$(NH_4)_3FeF_6$ structure	18, 115
$Ba(Co_{0.5}U_{0.5})O_3$	8.374			$(NH_4)_3FeF_6$ structure	17

TABLE 2.2 *(cont.)* $A^{2+}(B^{2+}_{0.5}B^{6+}_{0.5})O_3$ *(cont.)*

Compound	a (Å)	b (Å)	c (Å)	Remarks	References
$Ba(Co_{0.5}W_{0.5})O_3$	8.098			$(NH_4)_3FeF_6$ structure	15, 107
$Ba(Cr_{0.5}U_{0.5})O_3$	8.297			$(NH_4)_3FeF_6$ structure	17
$Ba(Cu_{0.5}U_{0.5})O_3$	8.18		8.84	tetragonal	17
$Ba(Fe_{0.5}Re_{0.5})O_3$	8.05			$(NH_4)_3FeF_6$ structure	18, 115
$Ba(Fe_{0.5}U_{0.5})O_3$	8.312			$(NH_4)_3FeF_6$ structure	17
$Ba(Fe_{0.5}W_{0.5})O_3$	8.133			$(NH_4)_3FeF_6$ structure	15
$Ba(Mg_{0.5}Os_{0.5})O_3$	8.08			$(NH_4)_3FeF_6$ structure	18
$Ba(Mg_{0.5}Re_{0.5})O_3$	8.082			$(NH_4)_3FeF_6$ structure	18, 115
$Ba(Mg_{0.5}Te_{0.5})O_3$	8.13			$(NH_4)_3FeF_6$ structure	108, 116
$Ba(Mg_{0.5}U_{0.5})O_3$	8.381			$(NH_4)_3FeF_6$ structure	17
$Ba(Mg_{0.5}W_{0.5})O_3$	8.099			$(NH_4)_3FeF_6$ structure	15, 16
$Ba(Mn_{0.5}Re_{0.5})O_3$	8.18			$(NH_4)_3FeF_6$ structure	18, 115
$Ba(Mn_{0.5}U_{0.5})O_3$	8.52			$(NH_4)_3FeF_6$ structure	17
$Ba(Ni_{0.5}Mo_{0.5})O_3$	4.0225				107
$Ba(Ni_{0.5}Re_{0.5})O_3$	8.04			$(NH_4)_3FeF_6$ structure	18, 115
$Ba(Ni_{0.5}U_{0.5})O_3$	8.336			$(NH_4)_3FeF_6$ structure	17
$Ba(Ni_{0.5}W_{0.5})O_3$	8.066			$(NH_4)_3FeF_6$ structure	15, 107
$Ba(Pb_{0.5}Mo_{0.5})O_3$					117
$Ba(Sr_{0.5}Os_{0.5})O_3$	8.43		8.72	tetragonal	18
$Ba(Sr_{0.5}Re_{0.5})O_3$	8.60		8.29	tetragonal	18, 115
$Ba(Sr_{0.5}U_{0.5})O_3$	8.84			$(NH_4)_3FeF_6$ structure	17
$Ba(Sr_{0.5}W_{0.5})O_3$	8.5			$(NH_4)_3FeF_6$ structure	16
$Ba(Zn_{0.5}Os_{0.5})O_3$	8.095			$(NH_4)_3FeF_6$ structure	18

TABLE 2.2 (cont.) $A^{2+}(B_{0.5}^{2+}B_{0.5}^{6+})O_3$ (cont.)

Compound	a (Å)	b (Å)	c (Å)	Remarks	References
$Ba(Zn_{0.5}Re_{0.5})O_3$	8.106			$(NH_4)_3FeF_6$ structure	18, 115
$Ba(Zn_{0.5}U_{0.5})O_3$	8.397			$(NH_4)_3FeF_6$ structure	17
$Ba(Zn_{0.5}W_{0.5})O_3$	8.116			$(NH_4)_3FeF_6$ structure	15
$Ca(Ca_{0.5}Os_{0.5})O_3$	5.73	7.87	5.80	orthorhombic	18
$Ca(Ca_{0.5}Re_{0.5})O_3$	5.67	8.05	5.78	orthorhombic	18
$Ca(Ca_{0.5}W_{0.5})O_3$	8.0			$(NH_4)_3FeF_6$ structure	16
$Ca(Cd_{0.5}Re_{0.5})O_3$	5.64	7.99	5.77	orthorhombic	18
$Ca(Co_{0.5}Os_{0.5})O_3$	5.47	7.70	5.59	orthorhombic	18
$Ca(Co_{0.5}Re_{0.5})O_3$	5.46	7.71	5.58	orthorhombic	18
$Ca(Fe_{0.5}Re_{0.5})O_3$	5.41	7.69	5.53	orthorhombic	18, 115
$Ca(Mg_{0.5}Re_{0.5})O_3$	5.48	7.77	5.56	orthorhombic	18
$Ca(Mg_{0.5}W_{0.5})O_3$	7.7			$(NH_4)_3FeF_6$ structure	16
$Ca(Mn_{0.5}Re_{0.5})O_3$	5.52	7.82	5.55	orthorhombic	18
$Ca(Ni_{0.5}Re_{0.5})O_3$	5.45	7.67	5.55	orthorhombic	18
$Ca(Sr_{0.5}W_{0.5})O_3$	8.1			$(NH_4)_3FeF_6$ structure	16
$Pb(Ca_{0.5}W_{0.5})O_3$					96
$Pb(Cd_{0.5}W_{0.5})O_3$	4.150	4.101	4.150	$\beta = 90°57'$	118
$Pb(Co_{0.5}W_{0.5})O_3$					119
$Pb(Mg_{0.5}Te_{0.5})O_3$	7.99			$(NH_4)_3FeF_6$ structure	116
$Pb(Mg_{0.5}W_{0.5})O_3$	4.0				95, 96
$Sr(Ca_{0.5}Mo_{0.5})O_3$					117
$Sr(Ca_{0.5}Os_{0.5})O_3$	8.21			$(NH_4)_3FeF_6$ structure	18
$Sr(Ca_{0.5}Re_{0.5})O_3$	5.76	8.21	5.85	orthorhombic	18
$Sr(Ca_{0.5}U_{0.5})O_3$	6.06	8.46	5.93	orthorhombic	17
$Sr(Ca_{0.5}W_{0.5})O_3$	8.2			$(NH_4)_3FeF_6$ structure	16
$Sr(Cd_{0.5}Re_{0.5})O_3$	5.73	8.16	5.81	orthorhombic	18
$Sr(Cd_{0.5}U_{0.5})O_3$	6.03	8.42	5.91	orthorhombic	17
$Sr(Co_{0.5}Mo_{0.5})O_3$	3.9367		3.9764		107, 113
$Sr(Co_{0.5}Os_{0.5})O_3$	7.86		7.92	tetragonal	18
$Sr(Co_{0.5}Re_{0.5})O_3$	7.88		7.98	tetragonal	18
$Sr(Co_{0.5}U_{0.5})O_3$	8.19			$(NH_4)_3FeF_6$ structure	17

TABLE 2.2 (cont.) $A^{2+}(B^{2+}_{0.5}B^{6+}_{0.5})O_3$ (cont.)

Compound	a (Å)	b (Å)	c (Å)	Remarks	References
$Sr(Co_{0.5}W_{0.5})O_3$	7.89		7.98	tetragonal	15, 113
$Sr(Cr_{0.5}U_{0.5})O_3$	8.09			$(NH_4)_3FeF_6$ structure	17
$Sr(Cu_{0.5}W_{0.5})O_3$	7.66		8.40	tetragonal	94
$Sr(Fe_{0.5}Os_{0.5})O_3$	7.85			$(NH_4)_3FeF_6$ structure	18
$Sr(Fe_{0.5}Re_{0.5})O_3$	7.86		7.89	tetragonal	18, 115
$Sr(Fe_{0.5}U_{0.5})O_3$	8.11			$(NH_4)_3FeF_6$ structure	17
$Sr(Fe_{0.5}W_{0.5})O_3$	7.96			$(NH_4)_3FeF_6$ structure	94
$Sr(Mg_{0.5}Mo_{0.5})O_3$					117
$Sr(Mg_{0.5}Os_{0.5})O_3$	7.86		7.92	tetragonal	18
$Sr(Mg_{0.5}Re_{0.5})O_3$	7.88		7.94	tetragonal	18
$Sr(Mg_{0.5}Te_{0.5})O_3$	7.94			$(NH_4)_3FeF_6$ structure	116
$Sr(Mg_{0.5}U_{0.5})O_3$	8.19			$(NH_4)_3FeF_6$ structure	17
$Sr(Mg_{0.5}W_{0.5})O_3$	7.9			$(NH_4)_3FeF_6$ structure	16
$Sr(Mn_{0.5}Re_{0.5})O_3$	8.01			$(NH_4)_3FeF_6$ structure	18
$Sr(Mn_{0.5}U_{0.5})O_3$	8.28			$(NH_4)_3FeF_6$ structure	17
$Sr(Mn_{0.5}W_{0.5})O_3$	8.01			$(NH_4)_3FeF_6$ structure	94
$Sr(Ni_{0.5}Mo_{0.5})O_3$	3.9237		3.9474	tetragonal	107, 113
$Sr(Ni_{0.5}Re_{0.5})O_3$	7.85		7.92	tetragonal	18
$Sr(Ni_{0.5}U_{0.5})O_3$	8.15			$(NH_4)_3FeF_6$ structure	17
$Sr(Ni_{0.5}W_{0.5})O_3$	7.86		7.91	tetragonal	15, 107, 113
$Sr(Pb_{0.5}Mo_{0.5})O_3$					117
$Sr(Sr_{0.5}Os_{0.5})O_3$	8.32		8.12	tetragonal	18
$Sr(Sr_{0.5}Re_{0.5})O_3$	8.41		8.13	tetragonal	18
$Sr(Sr_{0.5}U_{0.5})O_3$	6.22	8.65	6.01	orthorhombic	17
$Sr(Sr_{0.5}W_{0.5})O_3$	8.2			$(NH_4)_3FeF_6$ structure	16
$Sr(Zn_{0.5}Re_{0.5})O_3$	7.89		8.01	tetragonal	18
$Sr(Zn_{0.5}W_{0.5})O_3$	7.92		8.01	tetragonal	15, 113

TABLE 2.2 *(cont.)* $\quad A^{2+}(B^{1+}_{0.5}B^{7+}_{0.5})O_3$

Compound	a (Å)	b (Å)	c (Å)	Remarks	References
$Ba(Ag_{0.5}I_{0.5})O_3$	8.46			$(NH_4)_3FeF_6$ structure	108
$Ba(Li_{0.5}Os_{0.5})O_3$	8.100			$(NH_4)_3FeF_6$ structure	18
$Ba(Li_{0.5}Re_{0.5})O_3$	8.118			$(NH_4)_3FeF_6$ structure	18
$Ba(Na_{0.5}I_{0.5})O_3$	8.33			$(NH_4)_3FeF_6$ structure	18, 108
$Ba(Na_{0.5}Os_{0.5})O_3$	8.291			$(NH_4)_3FeF_6$ structure	18
$Ba(Na_{0.5}Re_{0.5})O_3$	8.296			$(NH_4)_3FeF_6$ structure	18
$Ca(Li_{0.5}Os_{0.5})O_3$	7.83			$(NH_4)_3FeF_6$ structure	18
$Ca(Li_{0.5}Re_{0.5})O_3$	7.83			$(NH_4)_3FeF_6$ structure	18
$Sr(Li_{0.5}Os_{0.5})O_3$	7.86			$(NH_4)_3FeF_6$ structure	18
$Sr(Li_{0.5}Re_{0.5})O_3$	7.87			$(NH_4)_3FeF_6$ structure	18
$Sr(Na_{0.5}Os_{0.5})O_3$	8.13			$(NH_4)_3FeF_6$ structure	18
$Sr(Na_{0.5}Re_{0.5})O_3$	8.13			$(NH_4)_3FeF_6$ structure	18

$$A^{3+}(B^{2+}_{0.5}B^{4+}_{0.5})O_3$$

Compound	a (Å)	b (Å)	c (Å)	Remarks	References
$La(Co_{0.5}Ir_{0.5})O_3$				orthorhombic	94
$La(Cu_{0.5}Ir_{0.5})O_3$				monoclinic	94
$La(Mg_{0.5}Ge_{0.5})O_3$	3.90				19
$La(Mg_{0.5}Ir_{0.5})O_3$	7.92				93
$La(Mg_{0.5}Nb_{0.5})O_3$					26, 94
$La(Mg_{0.5}Ru_{0.5})O_3$	7.91				26
$La(Mg_{0.5}Ti_{0.5})O_3$	3.932				19, 96
$La(Mn_{0.5}Ir_{0.5})O_3$	7.86			$(NH_4)_3FeF_6$ structure	26
$La(Mn_{0.5}Ru_{0.5})O_3$	7.84				26
$La(Ni_{0.5}Ir_{0.5})O_3$	7.90				26, 94
$La(Ni_{0.5}Ru_{0.5})O_3$	7.90				26
$La(Ni_{0.5}Ti_{0.5})O_3$	3.93				19
$La(Zn_{0.5}Ru_{0.5})O_3$	7.97				26
$Nd(Mg_{0.5}Ti_{0.5})O_3$	3.90				19

TABLE 2.2 *(cont.)*

$A^{2+}(B^{1+}_{0.25}B^{5+}_{0.75})O_3$

Compound	a (Å)	b (Å)	c (Å)	Remarks	References
Ba(Na$_{0.25}$Ta$_{0.75}$)O$_3$	4.137				27
Sr(Na$_{0.25}$Ta$_{0.75}$)O$_3$	4.055				27

$A(B^{3+}_{0.5}B^{4+}_{0.5})O_{2.75}$

Compound	a (Å)	b (Å)	c (Å)	Remarks	References
Ba(In$_{0.5}$U$_{0.5}$)O$_{2.75}$	8.551			(NH$_4$)$_3$FeF$_6$ structure	104

$A(B^{2+}_{0.5}B^{5+}_{0.5})O_{2.75}$

Compound	a (Å)	b (Å)	c (Å)	Remarks	References
Ba(Ba$_{0.5}$Ta$_{0.5}$)O$_{2.75}$	8.69			(NH$_4$)$_3$FeF$_6$ structure	24
Ba(Fe$_{0.5}$Mo$_{0.5}$)O$_{2.75}$	8.08			(NH$_4$)$_3$FeF$_6$ structure	28
Sr(Sr$_{0.5}$Ta$_{0.5}$)O$_{2.75}$	8.34			(NH$_4$)$_3$FeF$_6$ structure	24

2.3. Madelung Energy

The calculation of the binding energy of a crystal is one of the fundamental problems in the theory of solids. In this calculation, the basic assumption in the theory of cohesive energy of ionic crystals is that the solid can be considered as a system of positive and negative ions. In the NaCl structure, the shortest interionic distance is given by L. Each sodium ion is surrounded by 6 Cl$^-$ ions at a distance L, 12 Na$^+$ ions at a distance $L\sqrt{2}$, 8 Cl$^-$ ions at a distance $L\sqrt{3}$, etc. The Coulomb energy of the sodium ion in the field of all other ions is therefore,

$$U_M = -\frac{e^2}{L}\left(\frac{6}{\sqrt{1}} - \frac{12}{\sqrt{2}} + \frac{8}{\sqrt{3}} - \frac{6}{\sqrt{4}} + \frac{24}{\sqrt{5}}\right)$$

where e is the charge per ion. The coefficient of e^2/L is a pure number, determined only by the crystal structure and called the Madelung constant, M_L.

The Madelung constant for the cubic perovskite structure is 24.755. Using the equation

$$U_M = -M_L e^2/L$$

the Madelung energy for an assumed ideal structure for $CaTiO_3$ calculates out to be 4280.5.[117]

Rosenstein and Schor[121] determined the superlattice Madelung energy of idealized ordered cubic perovskites. These calculations were based on formulas $A_2^{2+}B^{4-n}B^{4+n}O_6$ which is twice the $A(B'_{0.5}B''_{0.5})O_3$ type formulas used throughout this report. In their formula $4 \pm n$ denoted a net charge of 4^+ plus $n\pm$ charges. The ordered structure for these compounds is shown in Fig. 2.5.

The method used for computing the change in Madelung energy due to charge ordering of the B ions is in accordance with the technique of superposition employed by Templeton,[122] since the ordering charges n^+ and n^- for an ordered perovskite were at alternate sites in a rocksalt structure. Using $M_L = 1.74756$ and $L = \frac{1}{4}$ supercell edge they derived the Madelung energy due to charge ordering,

$$U_M = -1.74756 \times 331.984 \times 0.5 \ n^2/L.$$

The Madelung energies are given in Table 2.3 below.

TABLE 2.3. *Table of Madelung Energies for Idealized Ordered Perovskite Compounds (after Rosenstein and Schor*[121]*)*

n	Compound	Supercell edge (Å)	Madelung energy ordering	Kcal mole^{-1} total
0	$Ca_2Ti_2O_6$	7.68	–	8561
1	$Ba_2Sc^{3+}Re^{5+}O_6$	8.16	142	8199
2	$Ba_2Ni^{2+}Re^{6+}O_6$	8.04	577	8755
3	$Ba_2Li^{1+}Re^{7+}O_6$	8.12	1286	9383

The results showed that the energy due to charge ordering was in some cases a significant fraction of the total Madelung energy especially where the charge difference in the B ions was large.

In a later study, Saltzman and Schor[123] calculated the Madelung energy of tetragonal perovskite structure. Madelung constants were determined for the 4–4, 3–5, and 1–7 tetragonal perovskite structures for axial ratios $\alpha = c/a$ varying from 0.90 to 1.10. Least-squares fits expressing the Madelung constant as a function of $(1-\alpha)$ also were reported. The results are given in Table 2.4.

TABLE 2.4. *Table of Madelung Energies (after Salzman and Schor[123])*

Compound	$\alpha = c/a$	Madelung constant	Madelung energy ordering	Total Madelung energy
Ba_2BaOsO_6	0.963	53.70	543	8235
Ba_2BaReO_6	0.963	53.70	544	8244
Ba_2SrReO_6	0.964	53.68	547	8289
Ba_2SrOsO_6	1.034	52.43	545	8260
Sr_2CoReO_6	1.013	52.78	587	8896
Sr_2FeReO_6	1.004	52.94	590	8945
Sr_2MgOsO_6	1.008	52.87	589	8934
Sr_2MgReO_6	1.008	57.87	588	8911
Sr_2NiReO_6	1.009	52.85	590	8941
Sr_2SrOsO_6	0.976	53.45	563	8531
Sr_2SrReO_6	0.967	53.62	559	8468
Sr_2ZnReO_6	1.015	52.74	585	8878

2.4. IONIC RADII

The ionic radii of the ions as given by Ahrens and as calculated in perovskite-type compounds are listed in Table 2.5. The ionic radii of the B ion was obtained in ABO_3-type compounds, while those of B′ and B″ were obtained in complex perovskite compounds. There are a number of ions which appear to differ considerably in radius in the structure of perovskite-type compounds as compared with those of Ahrens. In addition, the ionic radii of W^{5+} and Os^{7+} were calculated for the first time.

TABLE 2.5. *Ionic Radii*

	Ahrens	B†	B'‡	B''§
Ag^{1+}	1.26			
Au	1.37			
Cs	1.67			
Cu	0.96			
Fr	1.80			
I	0.62			
K	1.33			
Li	0.68			
Na	0.94			
Rb	1.47			
Tl	1.47			
Ag^{2+}	0.89			
Ba	1.34			
Be	0.35			
Ca	0.99		0.99	
Cd	0.97		0.97	
Co	0.73		0.73	
Cu	0.72			
Fe	0.74			
Ge	0.73			
Hg	1.10			
Mg	0.67		0.74	
Mn	0.80		0.80	
Ni	0.69		0.69	
Pb	1.20		1.20	
Pd	0.80			
Pt	0.80			
Ra	1.43			
Sn	0.93			
Sr	1.12		1.12	
Ti	0.76			
V	0.95			
Zn	0.74		0.74	
Ac^{3+}	1.18			
Al	0.55	0.558		
Am	1.07			
As	0.58			

TABLE 2.5 (cont.)

	Ahrens	B†	B'‡	B''§
Au	0.85			
B	0.23			
Bi	0.93			
Ce	1.07		1.06	
Co	0.63	0.56		
Cr	0.63	0.608		
Dy	0.92		0.94	
Er	0.89		0.91	
Eu	0.98		0.99	
Fe	0.64	0.628	0.63	
Ga	0.62	0.613		
Gd	0.97		0.97	
Ho	0.91		0.93	
In	0.81	0.714	0.78	
La	1.14		1.14	
Lu	0.85		0.86	
Mn	0.66	0.625	0.67	
Np	1.10			
Nd	1.04		1.04	
P	0.44			
Pa	1.13			
Pm	1.06			
Pr	1.06		1.02	
Pu	1.08			
Rh	0.68			
Sb	0.76			
Sc	0.81	0.686	0.74	
Sm	1.00		1.00	
Tb	0.93			
Ti	0.76	0.61		
Tl	0.95			
Tm	0.87		0.90	
V	0.74	0.625		
Y	0.92	0.773	0.92	
Yb	0.86		0.88	
Am^{4+}	0.92			
C	0.16			

TABLE 2.5 *(cont.)*

	Ahrens	B†	B′‡	B″§
Ce	0.94			
Ge	0.53			
Hf	0.78			
Ir	0.68			
Mn	0.60			
Mo	0.69			
Nb	0.74			
Np	0.95			
Os	0.69			
Pa	0.98			
Pb	0.84			
Pd	0.65			
Pr	0.92			
Pt	0.65			
Pu	0.93			
Rh	0.65			
Ru	0.67			
S	0.37			
Se	0.50			
Si	0.42			
Sn	0.71			
Tb	0.81			
Te	0.70			
Th	1.02			
Ti	0.68			
U	0.97			
V	0.63			
W	0.70			
Zr	0.79			
As^{5+}	0.46			
Bi	0.74			
Nb	0.69		0.69	
Os			0.67	
P	0.35			
Pa	0.89			
Re				0.68
Sb	0.62			

TABLE 2.5 (cont.)

	Ahrens	B†	B'‡	B''§
Ta	0.68			0.68
V	0.59			0.59
W				0.66
Cr^{6+}	0.52			
Mo	0.62			
Po	0.67			
Re	0.52			
S	0.30			
Se	0.42			
Te	0.56			
U	0.80			
W	0.62			0.62
At^{7+}	0.62			
Br	0.39			
Cl	0.27			
F	0.08			
I	0.50			
Mn	0.46			
Np	0.71			
Re	0.56			
Te	0.57			
Os				0.55

† Calculated for ABO_3-type compounds by S. Geller, *Acta Cryst.* **10**, 248 (1957).
‡ Ionic radii of B^{3+}: Calculated from complex perovskite compounds $A(B^{3+}_{0.5}Ta_{0.5})O_3$ by F. Galasso et al., UACRL D910269-5, Final Report, July 1965. Ionic radii of B^{2+}: Calculated from complex perovskite compounds $A(B^{2+}_{0.33}Ta_{0.67})O_3^{(23)}$.
§ Calculated from $A(B'_{0.5}B''_{0.5})O_3$ perovskite-type compounds.

References

1. H. D. MEGAW, *Proc. Phys. Soc.* **58,** 133, 326 (1946).
2. V. M. GOLDSCHMIDT, *Skrifter Norske Videnskaps-Akad. Oslo, I. Mat.-Naturv. Kl.*, No. 8 (1926).
3. R. W. G. WYCKOFF, *Crystal Structures* **2,** 359 (1964).
4. A. J. SMITH and A. J. E. WELCH, *Acta Cryst.* **13,** 653 (1960).
5. R. S. ROTH, *J. Research NBS*, RP 2736, **58,** (1957).
6. S. GELLER and A. E. WOOD, *Acta Cryst.* **9,** 563 (1956).
7. J. T. LOOBY and L. KATZ, *J. Am. Chem. Soc.* **76,** 6029 (1954).
8. W. F. DEJONG, *Z. Krist.* **81,** 314 (1932).
9. D. RIDGLEY and R. WARD, *J. Am. Chem. Soc.* **77,** 6132 (1955).
10. H. P. ROOKSBY, E. A. D. WHITE and S. A. LANGSTON, *J. Am. Ceram. Soc.* **48,** 447 (1965).
11. C. BRISI, *Ricerca Sci.* **24,** 1858 (1954).
12. G. H. JONKER, *Physica* **20,** 1118 (1954).
13. H. L. YAKEL, *Acta Cryst.* **8,** 394 (1955).
14. G. H. JONKER and J. H. VAN SANTEN, *Physica*, **16,** 337 (1950).
15. F. J. FRESIA, L. KATZ and R. WARD, *J. Am. Chem. Soc.* **81,** 4783 (1959).
16. E. G. STEWARD and H. P. ROOKSBY, *Acta Cryst.* **4,** 503 (1951).
17. A. W. SLEIGHT and R. WARD, *Inorg. Chem.* **1,** 790 (1962).
18. A. W. SLEIGHT, J. LONGO and R. WARD, *Inorg. Chem.* **1,** 245 (1962).
19. R. ROY, *J. Am. Ceram. Soc.* **27,** 581 (1954).
20. F. GALASSO, L. KATZ and R. WARD, *J. Am. Chem. Soc.* **81,** 820 (1959).
21. F. GALASSO, J. R. BARRANTE and L. KATZ, *J. Am. Chem. Soc.* **83,** 2830 (1961).
22. F. GALASSO and J. PYLE, *Inorg. Chem.* **2,** 482 (1963).
23. F. GALASSO and J. PYLE, *J. Phys. Chem.* **67,** 1561 (1963).
24. L. BRIXNER, *J. Am. Chem. Soc.* **80,** 3214 (1958).
25. F. GALASSO and W. DARBY, *J. Phys. Chem.* **66,** 131 (1962).
26. F. GALASSO and W. DARBY, *Inorg. Chem.* **4,** 71 (1965).
27. F. GALASSO and J. PINTO, *Inorg. Chem.* **4,** 255 (1965).
28. F. K. PATTERSON, C. W. MOELLER and R. WARD, *Inorg. Chem.* **2,** 196 (1963).
29. M. H. FRANCOMBE and B. LEWIS, *Acta Cryst.* **11,** 175 (1958).
30. W. H. ZACHARIASEN, *Skrifter Norske Videnskaps-Akad. Oslo,I. Mat.-Naturv. Kl.* No. 4 (1928).
31. I. NÁRAY-SZABÓ, *Műegyetemi Közlemények* **1,** 30 (1947).
32. R. P. OZEROV, N. V. RANNEV, V. I. PAKHOMOV, I. S. REZ and G. S. ZHDANOV, *Kristallografiya* **7,** 620 (1962).
33. I. NÁRAY-SZABÓ and A. KÁLMÁN, *Acta Cryst.* **14,** 791 (1961).
34. J. H. SMITH, *Nature* **115,** 334 (1925).
35. E. A. WOOD, *Acta Cryst.* **4,** 353 (1951).
36. L. L. QUILL, *Z. Anorg. Allgem. Chem.* **208,** 257 (1932).

37. P. VOUSDEN, *Acta Cryst.* **4**, 373 (1951).
38. M. WELLS and D. MEGAW, *Proc. Phys. Soc.* **78**, 1258 (1961).
39. H. F. KAY and J. L. MILES, *Acta Cryst.* **10**, 213 (1957).
40. D. SANTANA, *Anales Real. Soc. Españ. Fis. Quím. (Madrid)*, Ser. A, **44**, 557 (1948).
41. L. RIVOIR and M. ABBAD, *Anales Fís. Quím. (Madrid)*, **43**, 1051 (1947).
42. A. F. WELLS, *Structural Inorganic Chemistry*, Oxford University Press, Amen House, London (1950).
43. W. W. MALINOFSKY and H. KEDESDY, *J. Am. Chem. Soc.* **76**, 3090 (1954).
44. S. W. DERBYSHIRE, A. C. FRAKER and H. H. STADELMAIER, *Acta Cryst.* **14**, 1293 (1961).
45. L. H. BRIXNER, *J. Inorg. Nucl. Chem.* **14**, 225, (1960).
46. R. WEISS, *Compt. Rend.* **246**, 3073, (1958).
47. A. HOFFMAN, *Z. Physik. Chem.* **28**, 65 (1953).
48. L. E. RUSSELL, J. D. L. HARRISON and N. H. BRETT, *J. Nucl. Mater.* **2**, 310 (1960).
49. S. M. LANG, F. P. KNUDSEN, C. L. FILLMORE, R. S. ROTH, NBS Circ. 568 (1956).
50. I. NÁRAY-SZABÓ, *Naturwiss.* **31**, 466 (1943).
51. C. E. CURTIS, L. M. DONEY and J. R. JOHNSON, Oak Ridge Natl. Lab. ORNL–1681 (1954).
52. R. WARD, B. GUSHEE, W. MCCARROLL and D. H. RIDGELY, Univ. Conn. 2d Tech. Rep., NR-052–268, Contract ONR–367(00) (1953).
53. W. H. MCCARROLL, R. WARD and L. KATZ, *J. Am. Chem. Soc.* **78**, 2909 (1956).
54. L. W. COUGHANOUR, R. S. ROTH, S. MARZULLO and F. E. SENNETT, *J. Research NBS*, RP 2576, **54**, 149 (1955).
55. L. W. COUGHANOUR, R. S. ROTH, S. MARZULLO and F. E. SENNETT, *J. Research NBS*, RP 2580, **54**, 191 (1955).
56. J. BROUS, I. FANKUCHEN and E. BANKS, *Acta Cryst.* **6**, 67 (1953).
57. G. SHIRANE and R. PEPINSKY, *Phys. Rev.* **91**, 812 (1953).
58. B. JAFFE, R. S. ROTH and S. MARZULLO, *J. Research NBS*, RP 2626, **55**, 239 (1955).
59. J. J. RANDALL and R. WARD, *J. Am. Chem. Soc.* **81**, 2629 (1959).
60. W. W. COFFEEN, *J. Am. Ceram. Soc.* **36**, 207 (1953).
61. F. SUGAWARA and S. IIDA, *J. Phys. Soc. Japan* **20**, 1529 (1965).
62. W. H. ZACHARIASEN, *Acta Cryst.* **2**, 388 (1949).
63. A. RUGGIERO and R. FERRO, *Gazz. Chim. Ital.* **85**, 892 (1955).
64. M. L. KEITH and R. ROY, *Am. Mineralogist* **39**, 1 (1954).
65. A. WOLD and R. WARD, *J. Am. Chem. Soc.* **76**, 1029 (1954).
66. W. RÜDORFF and H. BECKER, *Z. Naturforsch.* **96**, 613 (1954).
67. J. A. W. DALZIEL and A. J. E. WELCH, *Acta Cryst.* **13**, 956 (1960).
68. R. C. VICKERY and A. KLANN, *J. Chem. Phys.* **27**, 1161 (1957).
69. S. GELLER and V. B. BALA, *Acta Cryst.* **9**, 1019 (1956).
70. V. S. FILIP'EV, N. P. SMOLYANINOV, E. G. FESENKO and I. N. BELYAEV, *Kristallografiya* **5**, 958 (1960).

71. A. I. ZASLAVSKII and A. G. TUTOV, *Dokl. Akad. Nauk SSSR* **135**, 815 (1960).
72. S. A. FEDULOV, Y. N. VENEVTSEV, G. S. ZHDANOV and E. G. SMAZHEVSKAYA, *Kristallografiya* **6**, 795 (1961).
73. A. RUGGIERO and R. FERRO, *Atti. Accad. Nazl. Lincei, Rend., Classe Sci. Fis. Mat. Nat.* **17**, 254 (1954).
74. S. GELLER, *Acta Cryst.* **10**, 243 (1957).
75. S. GELLER, J. *Chem. Phys.* **24**, 6 (1956).
76. A. WOLD, B. POST and E. BANKS, *J. Am. Chem. Soc.* **79**, 6365 (1957).
77. F. ASKHAM, I. FANKUCHEN, and R. WARD, *J. Am. Chem. Soc.* **72**, 3799 (1950).
78. W. C. KOEHLER and E. O. WOLLAN, *Phys. Chem. Solids* **2**, 100 (1957).
79. M. FOËX, A. MANCHERON and M. LINÉ, *Compt. Rend.* **250**, 3027 (1960).
80. D. D. KHANOLKAR, *Current Sci. (India)* **30**, 52 (1961).
81. M. KESTIGIAN and R. WARD, *J. Am. Chem. Soc.* **76**, 6027 (1954).
82. W. D. JOHNSON and D. SESTRICH, *J. Inorg. Nucl. Chem.* **20**, 32 (1961).
83. M. KESTIGIAN, J. G. DICKINSON and R. WARD, *J. Am. Chem. Soc.* **79**, 5598 (1957).
84. N. N. PADUROW and C. SCHUSTERIUS, *Ber. Deut. Keram. Ges.* **32**, 292 (1955).
85. L. JAHNBERG, S. ANDERSSON and A. MAGNÉLI, *Acta Chem. Scand.* **13**, 1248 (1959).
86. A. MAGNÉLI and R. NILSSON, *Acta Chem. Scand.* **4**, 398 (1950).
87. G. HÄGG, *Z. Physik. Chem.* **29** B, 192 (1935).
88. D. VAN DUYN, *Rec. Trav. Chim.* **61**, 669 (1942).
89. A. MAGNÉLI, *Arkiv. Kemi.* **1**, 269 (1949).
90. E. O. BRIMM, J. C. BRANTLEY, J. H. LORENZ and M. H. JELLINEK, *J. Am. Chem. Soc.* **73**, 5427 (1951).
91. B. W. BROWN and E. BANKS, *J. Am. Chem. Soc.*, **76**, 963 (1954).
92. M. ATOJI and R. E. RUNDLE, *J. Chem. Phys.* **32**, 627 (1960).
93. F. GALASSO, W. DARBY and J. PYLE, prepared at the United Aircraft Corporation Research Laboratories — not reported.
94. G. BLASSE, *J. Inorg. Nucl. Chem.* **27**, 993 (1965).
95. G. A. SMOLENSKII, A. I. AGRANOVSKAYA and V. A. ISUPOV, *Sov. Phys. Solid State* **1**, 907 (1959).
96. A. I. AGRANOVSKAYA, *Bulletin of Acad. Sciences of U.S.S.R. Physics Series* **24**, 1271 (1960).
97. F. GALASSO and J. PYLE, *J. Phys. Chem.* **67**, 533 (1963).
98. F. GALASSO and J. PINTO, *Nature* **207**, 70 (1965).
99. V. A. BOKOV and I. E. MYL'NIKOVA, *Sov. Phys., Solid State* **2**, 2428 (1961).
100. I. G. ISMAILZADE, *Sov. Phys., Cryst.* **5**, 292 (1960).
101. A. S. VISKOV, YU. N. VENEVTSEV and G. S. ZHDANOV, *Sov. Phys. Dokl.* **10**, 391 (1965).
102. L. BRIXNER, *J. Inorg. Nucl. Chem.* **15**, 352 (1960).

STRUCTURE OF PEROVSKITE-TYPE COMPOUNDS 49

103. C. KELLER, J. Inorg. Nucl. Chem. 27, 321 (1965).
104. A. W. SLEIGHT and R. WARD, Inorg. Chem. 1, 790 (1962).
105. F. GALASSO, G. LAYDEN and D. FLINCHBAUGH, J. Chem. Phys. 44, 2703 (1966).
106. V. S. FILIP'EV and E. G. FESENKO, Sov. Phys., Cryst. 6, 616 (1962).
107. L. BRIXNER, J. Phys. Chem. 64, 165 (1960).
108. A. W. SLEIGHT and R. WARD, Inorg. Chem. 3, 292 (1964).
109. V. S. FILIP'EV and E. G. FESENKO, Sov. Phys., Cryst. 10, 243 (1965).
110. G. A. SMOLENSKII, A. I. AGRANOVSKAYA, S. N. POPOV and V. A. ISUPOV, Sov. Phys., Tech. Phys. 3, 1981 (1958).
111. M. F. KUPRIYANOV and E. G. FESENKO, Sov. Phys. Cryst. 10, 189 (1965).
112. I. G. ISMAILZADE, Sov. Phys. Cryst. 4, 389 (1960).
113. M. F. KUPRIYANOV and E. G. FESENKO, Sov. Phys. Cryst. 7, 358 (1962).
114. G. BLASSE, Philips Research Reports 20, 327 (1965).
115. J. LONGO and R. WARD, J. Am. Chem. Soc. 83, 2816 (1961).
116. G. BAYER, J. Am. Ceram. Soc. 46, 604 (1963).
117. I. N. BELYAEV, L. I. MEDVEDEVA, E. G. FESENKO and M. F. KUPRIYANOV, Izv. Akad. Nauk SSSR Neorgan. Materialy, 1 6, (1965).
118. YU. E. ROGINSKAYA and YU. N. VENEVTSEV, Sov. Phys., Cryst. 10, 275 (1965).
119. YU. YA. TOMASHPOL'SKII and YU. N. VENEVTSEV, Sov. Phys. Solid State 6, 2388 (1965).
120. Q. C. JOHNSON and D. H. TEMPLETON, J. Chem. Phys. 34, 2004 (1961).
121. R. D. ROSENSTEIN and R. SCHOR, J. Chem. Phys. 1, 789 (1963).
122. D. H. TEMPLETON, J. Chem. Phys. 23, 1826 (1955).
123. M. N. SALTZMAN and R. SCHOR. J. Chem. Phys. 42, 3698 (1965).

CHAPTER 3

X-RAY DIFFRACTION AND ELECTRON PARAMAGNETIC STUDIES

3.1. X-ray Diffraction

In the previous chapter, the structures of perovskite-type compounds were described. This section of this chapter shows how X-ray diffraction can be used to determine unit cell data, in some cases, to determine the structure and to orient single crystals.

The X-ray diffraction powder patterns of compounds with the perovskite structure are those of a simple ~ 4 Å cubic cell with alternating weak and strong lines. The closer the A and B ions are in atomic number the weaker are those reflections for which $h+k+l \neq 2n$. Thus, the powder and single-crystal patterns of compounds with the "ideal" perovskite structure are extremely easy to identify. A Buerger precession photograph of $BaTiO_3$ taken with the beam parallel to the c-axis shows what a simple perovskite pattern looks like (see Fig. 3.1). Other perovskite-type compounds might produce patterns with slightly different distances between reflections and different relative intensities, but the basic pattern will be the same unless distortions or ordering exists in the structure.

The alignment of large single crystals for various measurements is also facilitated because of this simple structure.

When one axis is lengthened or shortened the structure exhibits tetragonal symmetry and the unique axis becomes "c". In the $BaTiO_3$ structure the distortion is small so that a splitting of reflections is observed best in those with higher indices. Reflections such as the 220, for example, split into the 202 and 220 (see Fig. 3.2 and Table 3.1). The X-ray pattern of $PbTiO_3$, where the distortion is greater, shows clear

TABLE 3.1. *X-Ray Data for* $BaTiO_3$

$d(Å)$	I	$h\ k\ l$
4.03	12	001
3.99	25	100
2.838	100	101
2.825	100	110
2.314	46	111
2.019	12	002
1.997	37	200
1.802	6	102
1.790	8	201
1.786	7	210
1.642	15	112
1.634	35	211
1.419	12	202
1.412	10	220
1.337	5	212
1.332	2	221
1.275	5	103
1.264	7	301
1.263	9	310
1.214	3	113

splitting of lower index reflections and it becomes more difficult to visualize the "ideal" cubic perovskite pattern from which it is derived.

When a and b are not changed, but the angle between them is other than 90°, an orthorhombic cell can be selected by taking the new a- and c-axis along the diagonals of the faces and taking a new b-axis perpendicular to the ac-plane. In the precession photographs shown in Fig. 3.3, it can be seen that the a'^* and b'^* axes in the $GdFeO_3$ photograph corresponding to the ~ 4 Å perovskite cell as in barium titanate are not at an angle of 90° to one another. However, the new axes a^* and c^*, which are located approximately midway between the small perovskite pseudocell axes, do form a right angle. Geller and Wood were able to index the powder pattern of $GdFeO_3$ using the data obtained from single-crystals studies (see Table 3.2). Note that the pattern for $GdFeO_3$

TABLE 3.2. *X-Ray Data for* $GdFeO_3$ *(after S. Geller and E. A. Wood,* Acta Cryst. *9, 562 (1950))*

$d(Å)$	I	$h\,k\,l$
3.880	M–S	110
3.831	VW	002
3.461	S	111
2.810	M–S	020
2.725	VS	112
2.677	M–S	200
2.636	M	021
2.304	W	103, 211
2.268	W–M	022
2.200	M	202
2.134	M	113
2.083	VW	122
1.936	M S	220
1.916	W–M	004
1.890	M	023
1.877	M	221
1.757	VW–W	213
1.741	VW	301
1.720	S	114, 131
1.700	VW	310
1.659	W	311
1.604	M	132
1.583	M	024
1.558	VW–W	204
1.553	S–VS	312
1.544	W–M	223
1.453	M–S	133
1.426	VW–W	115
1.416	VW	313
1.380	W–M	041
1.362	M–S	224
1.346	W	025
1.337	W–M	400
1.282	VW–W	142, 411
1.272	M–S	314, 331
1.266	VW–W	402
1.230	VW–W	412
1.226	M–S	241

TABLE 3.2 *(cont.)*

$d(\text{Å})$	I	$h\,k\,l$
1.223	W–M	332
1.214	S	116
1.207	W–M	420
1.202	W–M	225, 143
1.192	W–M	421
1.184	VW	324
1.182	VW–W	242
1.163	VW–W	026
1.459	VW–W	413
1.158	S	135
1.153	VW–W	206
1.152	S	333

is quite complex (see Fig. 3.2). This, of course, is caused by the inequalities in the cell parameters. The a^* of the monoclinic unit cell is approximately equal to the new $a/\sqrt{2}$, $c/\sqrt{2}$ and $b/2$, and these ratios are only approximately equal to each other. Thus, a split can be seen between the 110 and 002 reflections and further out between 024 and 204. As was pointed out in Chapter 2, a large number of perovskite-type compounds have the $GdFeO_3$ structure, and even some that do not can still be indexed on a similar unit cell or one which can be described by doubling cell edges. It should be noted that these distortions have been observed in neutron diffraction studies of a large number of these oxides.[1]

Some of the perovskite-type compounds also form a rhombohedral structure. Rhombohedral $LaAlO_3$, $NdAlO_3$ and $PrAlO_3$ probably have quite similar structures to that of $GdFeO_3$.

The structure of the complex perovskite compounds have not been so well characterized. If the two B ions are present in the perovskite structure in a random distribution, a simple perovskite pattern is obtained for the compound, such as the one shown in Fig. 3.4. However, most of the compounds of this type form ordered structures. When the two B ions are

TABLE 3.3. *X-Ray Data for* $Ba(Y_{0.5}Ta_{0.5})O_3$

$d(Å)$	I	hkl
4.88	40	111
4.24	10	200
2.98	100	220
2.543	30	311
2.430	20	222
2.108	70	400
1.930	10	331
1.875	< 10	420
1.722	80	422
1.624	20	511, 333
1.491	60	440
1.424	20	531
1.335	60	620
1.29	≪ 10	533
1.217	40	444
1.18	≪ 10	711, 551
1.127	70	642
1.10	≪ 10	731, 553
1.053	20	800
1.03	≪ 10	733
0.994	50	822, 660
0.974	≪ 10	751, 555
0.942	50	840
0.926	≪ 10	911, 753
0.899	40	664
0.885	≪ 10	931
0.860	50	844
0.848	≪ 10	933, 755, 771
0.827	70	10, 20, 862
0.815	≪ 10	951, 773
0.787	≪ 10	953

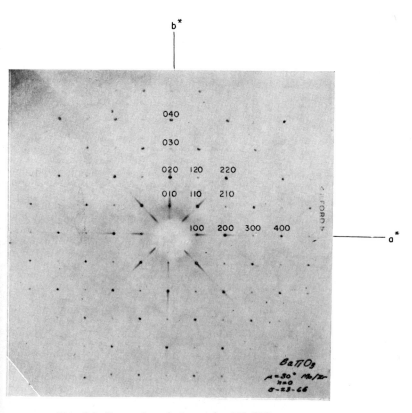

FIG. 3.1. Precession photograph of BaTiO$_3$.

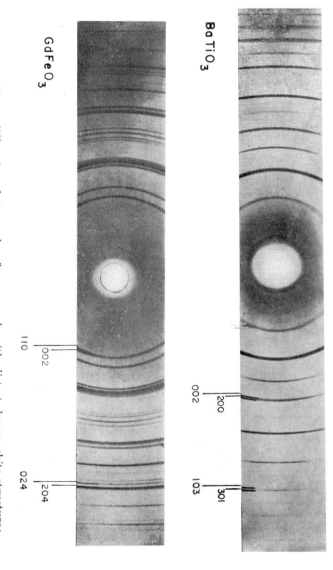

Fig. 3.2. X-ray diffraction photographs of compounds with distorted perovskite structure: Top—BaTiO$_3$, Cu radiation. Bottom—GdFeO$_3$, Co radiation.

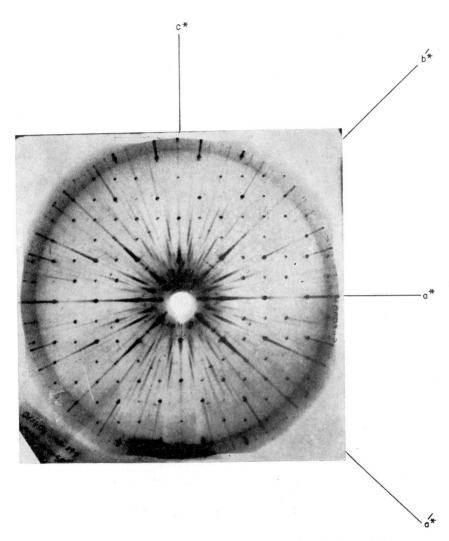

Fig. 3.3. Precession photograph of GdFeO$_3$ single crystal.

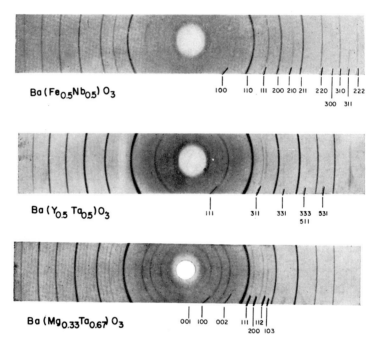

FIG. 3.4. X-ray powder diffraction photographs of unordered and ordered perovskite-type compounds:

Top—Ba(Fe$_{0.5}$Nb$_{0.5}$)O$_3$, Cu radiation.
Middle—Ba(Y$_{0.5}$Ta$_{0.5}$)O$_3$, Cu radiation.
Bottom—Ba(Mg$_{0.33}$Ta$_{0.67}$)O$_3$, Cu radiation.

present in equal quantities, the structure formed is the one described by Steward and Rooksby where B ions alternate at the corners of the unit cell to form a face-centered cubic lattice (see Fig. 2.5). The X-ray pattern of these compounds contain reflections for which h, k and l have to be all even or all odd, when the pattern is indexed on an ~ 8 Å cubic unit cell (see Fig. 3.4 and Table 3.3). The reflections such as the 111 and 311 for $Ba(Y_{0.5}Ta_{0.5})O_3$ are caused by ordering of the Y and Ta ions (see Fig. 3.4).

TABLE 3.4. *X-Ray Data for* $Ba(Mg_{0.33}Ta_{0.67})O_3$

d(Å)	I	$h\,k\,l$
7.040	20	001
5.006	30	100
4.100	30	101
2.887	100	102, 110
2.358	30	201, 003
2.045	60	202
1.669	80	212, 300, 104
1.444	60	204, 220
1.294	60	214
1.179	30	402
1.094	70	410
1.023	20	404
0.964	40	502
0.962	30	306

Since the intensity of any reflection in the powder pattern is given by the equation

$$I = |F|^2 \varrho \frac{(1+\cos^2 2\theta)}{\sin^2 \cos \theta}$$

and the multiplicity factor ϱ and the Bragg angle θ can be calculated for any line on the pattern, the relative intensity of a reflection can be determined if F, the structure factor, is known. For the ordering reflections, F is proportional to the difference in the atomic scattering factors or roughly equal to the difference in the atomic numbers of the B ions for small values of θ. Thus, ordered perovskite-type compounds which contain B ions with large differences in the atomic numbers will produce patterns with strong superstructure lines. A departure from perfect long-range order causes the superlattice lines to be weaker. Compounds in which the B ion charge or size differences are largest have structures with the highest degree of ordering. To date, no order–disorder phenomena with temperature have been reported for these compounds as is found in metal systems.

The cell sizes of compounds with the general formula $La(B^{2+}_{0.5}B^{4+}_{0.5})O_3$ appear to be cubic. Figure 3.5a shows a Buerger precession photograph of $La(Ni_{0.5}Ru_{0.5})O_3$. While the cell edge is ~ 8 Å, the lattice seems to be simple cubic and not face-centered cubic like the ordered perovskite-type compounds. These reflections must be caused by a movement of the ions from those in the ideal perovskite structure. For illustration purposes, the reflections which would be present if the B ions were ordered are marked on a diagram taken from a first-level precession photograph (see Fig. 3.5b). However, for this particular compound ordering of the B ions has not been proven.

When there is a tetragonal distortion of the ordered-perovskite structure, some authors report a unit cell with an ~ 8 Å a-axis with an ~ 8 Å c-axis and others select the body centered tetragonal cell by using the face diagonal of the small cubic cell as the a-axis. Similar axes are selected for orthorhombic distortions, thus the a and b cell edges are equal to $\sqrt{2}a$ and the c-axis is equal to $2a$ where $a \cong 4$ Å.

The ordered structure for compounds with the general formula $A(B'_{0.33}B''_{0.67})O_3$ is shown in Fig. 2.6. The X-ray patterns of these compounds have been indexed on a hexagonal cell of $a \cong 6$ Å and $c \cong 7$ Å. Table 3.4 presents the indexing of one of these compounds $Ba(Mg_{0.33}Ta_{0.67})O_3$ and Fig. 3.4c shows its X-ray photograph. While the hexagonal cell is the correct one, it is possible to index some of the patterns on a cubic cell of three times the small cubic a-axis. The intensities of the superstructure lines in this $Ba(Sr_{0.33}Ta_{0.67})O_3$ structure like those in the other are stronger for compounds which contain B ions with large differences in atomic number.

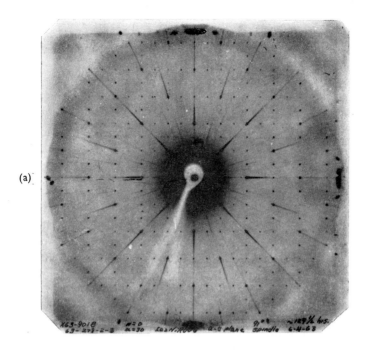

FIG. 3.5. (a) Precession photograph of La(Ni$_{0.5}$Ru$_{0.5}$)O$_3$ single crystal. (b) Diagram taken from first level precession photograph of La(Ni$_{0.5}$Ru$_{0.5}$)O$_3$. Circled reflections are the ones that would be present in an 8-Å ordered perovskite structure.

3.2. Electron Paramagnetic Resonance Studies

While X-ray diffraction measures an average effect in a structure, electron paramagnetic resonance is a powerful tool in studying the detailed atomic arrangement around a paramagnetic impurity ion. Danilyuk and Kharitonov[2] used EPR to study the centers responsible for the semiconducting properties of nonstoichiometric barium titanate. The samples were prepared by heating single crystals and ceramic discs for 40 min at 1200°C in a hydrogen atmosphere.

Electron paramagnetic resonance measurements were made at a frequency of 9400 Mc at 78°K. For the ceramic disc and single crystal an electron paramagnetic spectrum was obtained which consisted of six equidistant components of the same intensity and a slightly asymmetric line in the center of the spectrum.

It was postulated that the observed spectrum was due to an electron captured by the oxygen vacancy center which was localized in the region of Ti^{4+} cation and which can be considered to be a Ti^{3+} ion. However, the electron had a greater probability of being next to an oxygen vacancy and there was only partial localization next to the Ti^{4+}.

Sakudo et al.[3] studied the effect of a d.c. electric field on the resonance spectra of $BaTiO_3$ near the Curie temperature using Fe^{3+} as an impurity ion. The temperature of the microwave resonant cavity was carefully controlled and the resonance spectra were observed with and without the d.c. electric field. As would be expected from X-ray studies, the relative intensity of the lines due to the tetragonal phase increased by the application of the bias field and that due to the cubic phase decreased correspondingly.

Sakudo and Unoki[4] examined by (EPR) the behavior of $BaTiO_3$ in the neighborhood of the Curie temperature to determine if a fluctuation of the polarization might exist. The temperature was carefully controlled. However, no evidence of a resonance spectrum between those of the tetragonal and cubic phase existed. Furthermore, "a" was essentially constant between 120°C and 140°C.

Rimai and deMars[5] found that 0.2 atom % Gd^{3+} substituted into the barium titanate structure produces seven lines in a room-temperature electron spin resonance spectrum which they interpreted as Gd^{3+} substituting for barium in a tetragonal site. Takeda and Watanabe[6] in a similar study found additional lines in the electron paramagnetic resonance spectrum which they attributed to some Gd^{3+} substitution in the titanium site as well, producing a cubic site. Substitution of larger ions in the B sites has been well established by X-ray diffraction studies.

Electron paramagnetic resonance studies of Fe^{3+} in strontium titanate were conducted by Dobrov et al.[7] down to 1.9°K, feeling that a sharp ferroelectric phase transition should be observed at about 40°K. Instead, a smooth transition from the nonferroelectric state to a ferroelectric state was observed.

Rimai and deMars[5] also observed a cubic to tetragonal phase transformation in $SrTiO_3$ using a Gd^{3+}-doped sample in electron paramagnetic studies. The transition temperature was 110°K. Like Dobrov et al., Rimai found a smooth transition contrary to the corresponding case in $BaTiO_3$ where the change was sharp, leaving some doubt as to whether $SrTiO_3$ is a ferroelectric material.

Kirkpatrick et al.[8] observed an axial electron paramagnetic resonance spectrum in iron-doped strontium titanate. The same spectrum was observed after charge displacement due to heat treatment in the dark or by reduction in $SrTiO_3$ which contained other transition metals in addition to iron. These experiments indicate that charge compensation occurred at a nearest oxygen site, and this charge compensation was produced by an oxygen vacancy next to a substitutional Fe^{3+} ion. In addition, the axial Fe^{3+} spectrum after heat treatment was most likely due to a hole trapped at a neutral complex containing an Fe^{2+} ion and an O^{2-} vacancy.

Rimai et al.[9] investigated the electron paramagnetic resonance spectra of Fe^{3+}, Gd^{3+}, Eu^{3+} and Cr^{3+} in cubic $SrTiO_3$ as a function of hydrostatic pressure to 10,000 bars and as a function of temperature between 300 and 110°K. These studies indicated that the local compressibility about the Cr^{3+} and Fe^{3+} sites were about twice the bulk value. In addition, the isoelectronic ions Gd^{3+} and Eu^{3+} showed a difference in the temperature dependence of the ground-state cubic field parameter b_0, which could be attributed to the polarization of the lattice by uncompensated charge of Gd^{3+}.

REFERENCES

1. W. C. KOEHLER and E. O. WOLHAM, *J. Phys. Chem. Sol.* **2**, 100 (1957).
2. Y. L. DANILYUK and E. V. KHARITONOV, *Sov. Phys. Solid State* **6**, 260 (1964).
3. T. SAKUDO, H. UNOKI and S. MAEKAWA, *J. Phys. Soc. Japan* **18**, 913 (1963).
4. T. SAKUDO and H. UNOKI, *J. Phys. Soc. Japan* **19**, 2109 (1964).
5. L. RIMAI and G. A. DEMARS, *Phys. Rev.* **127**, 702 (1962).
6. T. TAKEDA and A. WATANABE, *J. Phys. Soc. Japan* **19**, 1742 (1964).
7. W. I. DOBROV, R. F. VIETH and M. E. BROWNE, *Phys. Rev.* **115**, 79 (1959).
8. E. S. KIRKPATRICK, K. A. MÜLLER and R. S. RUBINS, *Phys. Rev.* **135**, A86 (1964).
9. L. RIMAI, T. DEUTSCH and B. D. SILVERMAN, *Phys. Rev.* **133**, A1123 (1964).

CHAPTER 4
CONDUCTIVITY

MANY of the perovskites are noted for their high electrical resistivities, which make them useful as dielectric materials. However, some of the perovskite-type phases such as $CaMoO_3$, $SrMoO_3$, $LaTiO_3$ and $LaVO_3$, which contain B ions in lower than their most stable oxidation state, and

$La_{1-x}Sr_xMnO_3$, $SrTiO_{3-x}$, $SrVO_{3-x}$ and $Ba_{1-x}La_xTiO_3$,

which contain the B ions in two valence states, are considered to be fairly good conductors or semiconductors. Probably the best conductors are the tungsten bronzes with the cubic perovskite structure.

4.1. CONDUCTORS

Straumanis and Dravnieks[1] and Huibregtse et al.[2] have shown that the cubic sodium tungsten bronzes (Na_xWO_3, x between 0.3 and 0.95) are highly conducting and exhibit a positive temperature coefficient of resistance. Resistance measurements made by Brown and Banks[3] over a temperature range of $-160°C$ to $360°C$ for single crystals of six different compositions showed a linear dependence of resistivity over the entire range (Fig. 4.1), indicating that the nature of conduction in these bronzes is metallic. In addition, a series of resistivity isotherms as a function of bronze composition showed that a minimum in resistivity existed in a phase with 0.7 mole % sodium (see Fig. 4.2). An interpretation of this minimum was presented in terms of equilibrium between undissociated sodium atoms and sodium ions plus free electrons. For sodium concentrations below $x = 0.70$ each atom introduced one free electron and one random scattering center (Na^+), while above $x = 0.70$, the addition of undissociated sodium atoms contributed only additional scattering centers.

Similar data by Gardner and Danielson[4] also showed a minimum resistivity for Na_xWO_3 phase where $x = 0.75$ in good agreement with Brown et al. However, the Hall coefficient had a value which corresponded to one free electron

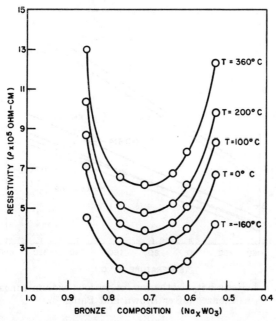

FIG. 4.1. Plot of resistivity vs. temperature for six compositions of sodium tungsten bronzes Na_xWO_3 (after Brown and Banks[3]).

for each sodium atom even for values of x greater than 0.75. Therefore they did not agree with Banks et al. as to their reason for the minimum in resistivity, but instead proposed that ordering of the sodium atoms at $x = 0.75$ might be the cause.

In a later study Ellerbeck et al.[5] cast doubt on this postulation by selecting crystals which were electrically homogeneous and found no minimum in the resistivity versus sodium concentration curve, but instead found that the electrical conductivity increased approximately linearly with sodium content at room temperature. These results indicated that large macroscopic growth defects might be an important factor in producing the minimum in the resistivity of the crystals observed by Brown and Gardner. However, Atoji and Rundle[6] did find ordering of the sodium atoms in $Na_{0.75}WO_3$ as was suggested by Gardner *et al.* Mackintosh[7] in evaluat-

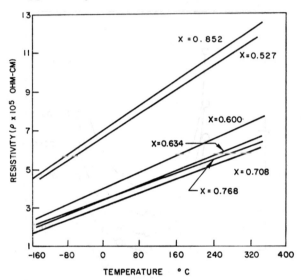

FIG. 4.2. Resistivity isotherms for various sodium tungsten bronzes (after Brown and Banks[3]).

ing these various studies felt that the minimum in resistivity should exist at this composition. It is probable that ordering in $Na_{0.75}WO_3$ depends on the method of sample preparation, which would account for the variation in results.

Sienko and Truong[8] made resistivity measurements on $Li_{0.394}WO_3$, $Li_{0.377}WO_3$ and $Li_{0.365}WO_3$ and found that the conductivity in these phases also is metallic from -150 to $90°$. It can be seen in Fig. 4.3 that the electrical resistance increases linearly with increasing temperature. As in the case of sodium tungsten bronzes there was one free electron carrier per alkali atom.

Thus, the metallic conduction observed in both the lithium and sodium tungsten bronzes (A_xWO_3) with the perovskite structure has been attributed to x electrons, one for each sodium or lithium atom, in a partially filled band of electron states. It has been proposed that this conduction band in the sodium bronze is produced by (1) Na–Na bonding from the overlapping of sodium $3p$ orbitals,[7] (2) W–W bonding from

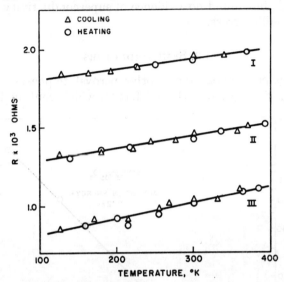

FIG. 4.3. Electrical resistance vs. absolute temperature of single crystals of Li_xWO_3; curve I, $Li_{0.394}WO_3$, curve II, $Li_{0.377}WO_3$, curve III, $Li_{0.375}WO_3$ (after Sienko et al.[8]).

the overlapping of tungsten T_{2g} orbitals[9] or (3) covalent bonding from a mixing of the oxygen $p\pi$ orbitals and tungsten T_{2g}.[10] Ferretti et al.,[11] after studying some perovskite-type compounds, felt that proposal three is probably the correct one.

4.2. SUPERCONDUCTORS

Sweedler et al.[12] investigated superconductivity in the tungsten bronzes. While sodium, potassium, rubidium and cesium tungsten bronzes were found to be superconductors, no perovskite bronzes were observed to be superconductors.

Schooley and co-workers[13] measured superconducting transitions in three samples of reduced strontium titanate. The reduced crystals were prepared by prolonged heating in vacuum of 10^{-5} to 10^{-7} mm Hg. The transitions occurred at 0.25°K and 0.28°K, respectively. Superconductivity has also been observed in reduced phases in the systems $(Ba_xSr_{1-x})TiO_3$ and $(Ca_ySr_{1-y})TiO_3$ when $x \leq 0.1$ and $y \leq 0.3$.[7]

Sweedler[14] also made measurements on $Sr_{0.82}NbO_3$ and $KTaO_{3-x}$ but found no evidence of superconductivity down to ~ 30 millidegrees.

4.3. SEMICONDUCTORS

As was pointed out above, other perovskite phases such as $CaMoO_3$, $SrMoO_3$, $LaTiO_3$ and $LaVO_3$ which contain B ions

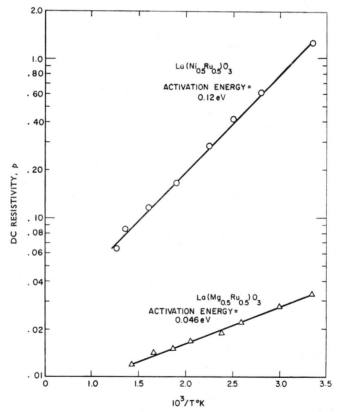

FIG. 4.4. Log resistivity vs. 1000/T for $La(Ni_{0.5}Ru_{0.5})O_3$ and $La(Mg_{0.5}Ru_{0.5})O_3$ (after Galasso and Darby[16]).

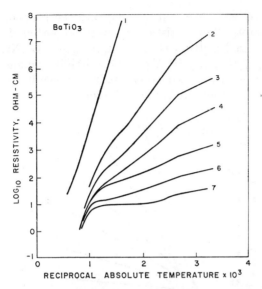

Fig. 4.5. Log resistivity vs. reciprocal temperature for $BaTiO_{3-x}$ phases (after Weise and Lesk[17]).

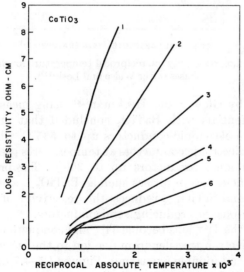

Fig. 4.6. Log resistivity vs. reciprocal temperature for $CaTiO_{3-x}$ phases (after Weise and Lesk.[17])

with electrons in the d-shells may also be semiconductors. For example, low resistances were reported for single crystals of $La(Ni_{0.5}Ru_{0.5})O_3$, $La(Mg_{0.5}Ru_{0.5})O_3$, $La(Ni_{0.5}Ir_{0.5})O_3$ and $La(Zn_{0.5}Ru_{0.5})O_3$.[15] A plot of resistivity versus temperature for the former two compounds is shown in Fig 4.4 and indicates that they are semiconductors.

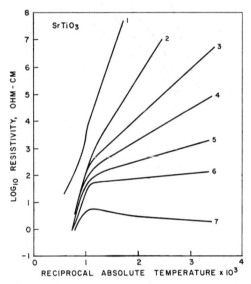

FIG. 4.7. Log resistivity vs. reciprocal temperature for $SrTiO_{3-x}$ phases (after Weise and Lesk[17]).

Studies by Glower and Heckman[16] using measurements of cell potentials with $BaTiO_3$ concluded that pure single crystals are electronic conductors up to 535°C, while ceramics were electronic conductors at temperatures greater than 500°C and ionic conductors up to 300°C. The presence of oxygen deficiencies in phases such as $BaTiO_{3-x}$ as well as in $CaTiO_{3-x}$ and $SrTiO_{3-x}$ reduces the resistivity of the stoichiometric phases producing semiconductors.

Some of the Ti^{4+} ions become Ti^{3+} and conduction can take place by electron transfer from one ion to the other through an oxygen ion (double exchange). Figures 4.5–4.7 show families of curves for each of the alkaline earth titanates,[17] each sample of which was more strongly reduced than the precedng one. Table 4.1 gives a tabulation of the band gaps, activation energies and percents of reduction.

The reduced samples of each compound formed a family of straight lines in the lower temperature range. Discontinuities in slopes for the reduced $BaTiO_3$ samples occurred

TABLE 4.1. *Data Obtained from Curves in Figs. 4.5–4.7 (after Weise and Lesk* [17])

	Curve no.	Reduction weight %	Act energy electron volts
$CaTiO_3$	1	0	2.61
	2	–	1.19
	3	–	0.651
	4	–	0.426
	5	–	0.433
	6	–	0.286
	7	0.06	0.086
$SrTiO_3$	1	0	3.27
	2	–	1.24
	3	–	0.716
	4	–	0.467
	5	–	0.221
	6	–	0.087
	7	0.23	–
$BaTiO_3$ above Curie temp.	1	0	2.47
	2	–	1.12
	3	–	0.858
	4	–	0.765
	5	–	0.389
	6	–	0.337
	7	0.1	0.298
$BaTiO_3$ below Curie temp.	1	0	
	2	–	0.516
	3	–	0.387
	4	–	0.349
	5	–	0.214
	6	–	0.151
	7	0.1	0.119

at the temperature where a change in its crystal structure occurs, about 120°C.

A similar change in oxidation state of the B ion can be accomplished by adding a foreign ion such as shown below[18] (controlled valency) in Table 4.2.

TABLE 4.2. *Examples of Controlled-valency Semiconductors (after Verwey et al.* [18])

Compound	Oxide added	Normal ion present	New ions in system
$CaTiO_3$	La_2O_3	Ti^{4+}	Ti^{3+}
$SrTiO_3$	La_2O_3	Ti^{4+}	Ti^{3+}
$BaTiO_3$	La_2O_3	Ti^{4+}	Ti^{3+}
$CaMn^{4+}O_3$	La_2O_3	Mn^{4+}	Mn^{3+}
$LaMn^{3+}O_3$	CaO	Mn^{3+}	Mn^{4+}
$LaMn^{3+}O_3$	SrO	Mn^{3+}	Mn^{4+}
$LaFe^{3+}O_3$	SrO	Fe^{3+}	Fe^{4+}

If the substituted ion has a lower valency the result is a p-type semiconductor; with foreign ions of higher valency n-type semiconductors are produced.

One of the problems of measuring the conductivity in these semiconductors is they often have to be used as powder compacts which have porosity and may have differences in composition on the surface of the crystallites. Thus the measurements may often depend on voltage and on frequency. Another problem is often the contact resistance between samples and electrodes. If a two-electrode system is used, fired on silver or many other electrode materials can be employed for p-type oxides; however, only a few types of electrodes can be used for n-type oxides. Graphite, electroless nickel (plating nickel from solution) and In-amalgams have proved to be satisfactory. Of course, whenever possible the four-probe system should be employed to minimize the errors involved from contact resistances.

Once the contact problem is solved a plot of log resistivity versus $1/T$ usually produces a straight line from which the activation energy is obtained. However, many of the perovskite-type compounds undergo phase transformations or are magnetic which influences their conduction. In a plot of log σ versus $1/T$ (Fig. 4.8) for $La_{0.8}Sr_{0.2}MnO_3$, for example, it can be seen that below the Curie temperature a minimum occurs and then above this temperature a much lower resistance appears. The electron transfers much easier if the spin moments are parallel. Ferromagnetic (La, Sr)CoO$_3$ exhibits nearly metallic conductivity,[19] while antiferromagnetic (La, Sr)FeO$_3$ and (La, Sr)CrO$_3$ have a much higher resistivity.[20]

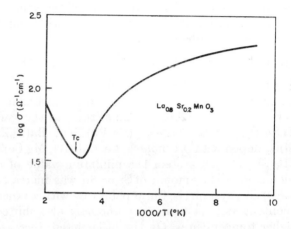

FIG. 4.8. Conductivity of $La_{0.8}Sr_{0.2}MnO_3$ as a function of reciprocal temperature (arrow indicates Curie point) (after Jonker and Van Stanten[19]).

Probably the most widely studied materials of the controlled valency type are the rare-earth-doped barium titanate-type polycrystalline bodies. The resistivity of barium titanate which is extremely high is reduced by the substitution of certain elements. Haaijman et al.[21] found that these phases thus treated are bluish and have resistivities lower than 10^3 ohm-cm which suddenly increase with temperature near the Curie point. Later studies were conducted by Sauer and Flaschen[22] and by Harman.[23] The former scientists described the materials as positive temperature coefficient of resistance thermistor materials. Harman studied the effect of adding samarium to barium titanate. Saburi[24] added Sm and also small amounts of Bi, Ce, La, Nb, Pr, Sb and Ta to different samples of barium titanate and found that they each reduced the resistance and caused the samples to exhibit an anomalous positive temperature coefficient of resistivity. However, doped barium titanate single crystals did not show a similar anomaly in resistivity reported for these polycrystalline samples. Thermo-e.m.f. for some representative polycrystalline bodies gave values of 690 to 869 $\mu V/°C$ and the sign of the generated e.m.f. indicated that they were n-type semiconductors. The bluish color of these materials suggested the presence of Ti^{3+} ions, and Saburi thought the following process of substitution occurred,

$$BaTi^{4+}O_3 + xLa^{3+} = Ba_{1-x}La_x^{3+}(Ti_{1-x}^{4+}Ti_x^{3+})O_3$$

producing the electrons for conduction.

In a later study, Saburi[25] investigated the resistivity and crystal structures for different compositions in the systems $Ba,Sr(Ti,Sn)O_3$, $(Ba,Ca,Sr)TiO_3$, $(Ba,Pb)TiO_3$, $Ba(Ti,Zr)O_3$, $Ba(Ti,Si)O_3$ doped with 0.1 mole % Ce and $(Ba, Mg,Ce)TiO_3$. In the $Ba,Sr(Ti,Sn)O_3$ system the minimum value of resistivity increased as the amount of Sr or Sn was increased. In the $(Ba,Pb)TiO_3$ system the Curie point rose with an increase in lead content and the resistivity anomaly also shifted toward higher temperatures. On the other hand, increases in the amount of Zr in the $Ba(Ti,Zr)O_3$ shifted the Curie point and the resistivity anomaly to lower temperatures.

Tennery and Cook[26] studied the effect of adding 0.00015 to 0.0030 mole fraction of rare earth oxides on the resistivity of $Ba(Ti,Zr)O_3$. When samarium, gadolinium or holmium was added in increasing amounts the resistivity dropped, passed through a minimum and rose again (see Fig. 4.9). Tennery et al. felt that the concentration at which the minimum occurred was related to the ionic radii of the added ion. The $BaTiO_3$ was n-type in the tetragonal phase and p-type in the cubic phase indicating that there was probably a mobility decrease in the major carrier at the Curie temperature because of a narrowing of the conduction band associated with the tetragonal–cubic transformation.

Peria et al.[27] also tried to explain the cause of the positive temperature coefficient (PTC) effect. The data available were reconstructed and the fact that when the resistivities of the samples were measured at high frequencies the PTC was not observed was weighed heavily. They felt that the strain

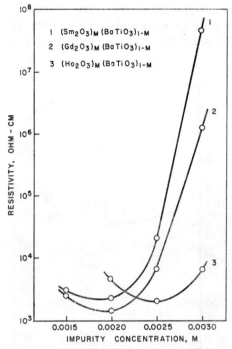

FIG. 4.9. Effect of Sm^{3+}, Gd^{3+} and Ho^{3+} concentration on the resistivity of $BaTiO_3$ (after Tennery and Cook[26]).

caused by the rapid change that begins to occur 20–30°C below the Curie point was responsible for a change in contact resistance between ceramic particles with temperature. Heywang[28] thought the PTC was caused by a change in height of blocking barriers set up across grains by electrons trapped at surface states.

Ryan and Subbarao[29] measured the Hall constant of lanthanum-doped $BaTiO_3$ as a function of temperature and felt that the carriers were being lost in the intergranular region of the ceramic near the Curie temperature. The experimental data, however, did not help to make a choice between the models of Peria and Heywang. Goodman elucidated the cause of the conduction anomaly of doped barium titanate even further by measuring the resistivity of both single crystals and ceramics of samarium-doped $BaTiO_3$.[30] The PTC anomaly was not present in the single crystal, but was present in a ceramic pressed and sintered from materials produced by crushing the single crystal (see Fig. 4.10). The

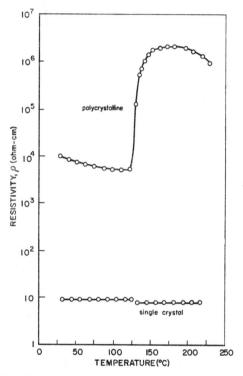

FIG. 4.10. Resistivity vs. temperature for samarium-doped barium titanate—polycrystalline and single crystal samples (after Goodman[22]).

grain-boundary region was visualized as an electron exhaustion layer 10^2 to 10^4 Å thick with a much higher resistivity than the inside of the grain. However, even after these studies a choice between the various models proposed for the PTC effect could not be made.

Because of the possible application of these materials as thermistors, many investigations were made to produce material with reproducible properties. Sauer and Fisher[31] found that mixing $La_2(C_2O_4)_2 \cdot 9H_2O$, $BaCO_3$, $SrCO_3$ and TiO_2, in appropriate proportions and mulling in a controlled manner, pressing into a pellet form and firing at 1375° resulted in reproducible samples. Gallagher et al.[32] used a technique for forming semiconducting barium titanate which involved coprecipitating the lanthanum in the preparation of barium titanyl oxalate and precipitation of lanthanum hydroxide in a slurry of the titanate. In a later study, Gallagher and Schey[33] added to the knowledge of sample preparation by studying the thermal decomposition of some substituted barium titanyl oxalates. All of these investigations indicated that reproducible thermistor samples can be prepared only with careful control of the preparation conditions adding credence to the postulation that the PTC effect is not a property of the single crystal material itself.

4.4. Thermoelectricity

The figure of merit (Z) of a thermoelectric material is defined by the expression

$$Z = \frac{S^2}{\varrho \varkappa}$$

where S is the Seebeck coefficient ($\mu V/\deg$), ϱ the electrical resistivity (ohm-cm) and $\varkappa$ the thermal conductivity (watts/cm deg). Thus a good thermoelectric material should have a high Seebeck coefficient, low electrical resistivity, low thermal conductivity and good physical and chemical stability at high temperatures. The most important group of thermoelectric materials are semiconductor compounds. Materials such as Bi_2Te_3 and $PbTe$ have been prepared with figures of merit as high as 3×10^{-3} at 25°C and 1.3×10^{-3} at 450°C, respectively.

A number of laboratories, in searching for better materials, conducted studies on perovskite-type compounds for ther-

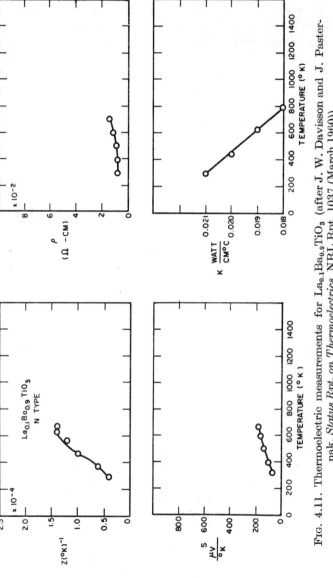

FIG. 4.11. Thermoelectric measurements for $La_{0.1}Ba_{0.9}TiO_3$ (after J. W. Davisson and J. Pasternak, *Status Rpt. on Thermoelectrics*, NRL Rpt. 1037 (March 1960)).

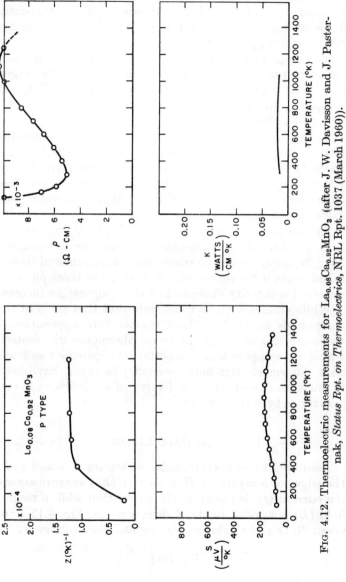

FIG. 4.12. Thermoelectric measurements for $La_{0.08}Ca_{0.92}MnO_3$ (after J. W. Davisson and J. Pasternak, Status Rpt. on Thermoelectrics, NRL Rpt. 1037 (March 1960)).

moelectric applications. It was felt that in "controlled-valency" perovskites, it would be easy to control the carrier concentration, the starting materials would be inexpensive, stable at higher temperatures, probably would suffer little radiation damage and trace impurities would not be important. A large number of phases were investigated and some of the more promising materials are described below. Figures 4.11 and 4.12 present data for $La_{0.1}Ba_{0.9}TiO_3$ and $La_{0.08}Ca_{0.92}Mn_3$ which appear to be good thermoelectrics at higher temperatures.[34]

Studies by General Ceramics[35] on several compositions with the formula $La_{1-x}Sr_xFeO_3$ showed that the Seebeck coefficient (p-type) was very low at room temperature, but rose with temperature and the electrical resistance of these phases decreased with temperature. Thus the high-temperature properties looked promising.

The Seebeck coefficient (n-type) of $SrMnO_3$ was found to decline with rising temperature, while the coefficient of $La_{1-x}Sr_xMnO_3$ phases increased to a maximum and then declined as the temperature was increased. For these phases the electrical resistivity decreased as the temperature increased.

While studies such as these indicated that some of these compounds might have high temperature application, researchers still have not produced thermoelectric materials which can compete with the better intermetallic thermoelectrics. Thermoelectric measurements, however, have helped to understand better the mechanism of conductivity in many perovskite type compounds.

4.5. Hall Effect

Assuming three perpendicular directions x, y and z along the edges of a crystal, a Hall voltage (Eh) appears across x, if a current (Ic) is passed in the y direction while a magnetic field (B) is applied in the Z direction (see Fig. 4.13). For a given B, Ic and thickness, the no-load Hall voltage is:

$$Eh = Rh\left(\frac{IcB}{t}\right)$$

where Rh is the Hall constant.

While the use of the Hall effect has found a number of applications for some semiconductor materials, the mobility of the carriers in semiconducting perovskites is so small as to make the use of these materials impractical for applications. The mobility in materials with large energy gaps is of

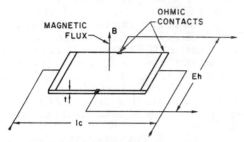

FIG. 4.13. Hall measurement.

the order 1 to 50 (cm/sec)/(volt/cm) as compared with good semiconductors which have mobilities of 500 to 5000 (cm/sec)/(volt/cm). The Hall effect, however, has been used with the thermoelectric effect to elucidate the mechanism of conduction in many perovskite-type compounds.

REFERENCES

1. M. E. STRAUMANIS and A. DRAVNIEKS, *J. Am. Chem. Soc.* **71**, 683 (1949).
2. E. J. HUIBREGTSE, D. E. BARKER and G. C. DANIELSON, *Phys. Rev.* **82**, 770 (1951).
3. B. W. BROWN and E. BANKS, *Phys. Rev.* **84**, 609 (1951).
4. W. R. GARDNER and G. C. DANIELSON, *Phys. Rev.* **93**, 46 (1954).
5. L. D. ELLERBECK, H. R. SHANKS, P. H. SIDLES and G. D. DANIELSON, *J. Chem. Phys.* **35**, 298 (1961).
6. M. ATOJI and R. E. RUNDLE, *J. Chem. Phys.* **32**, 627 (1960).
7. A. R. MACKINTOSH, *J. Chem. Phys.* **38**, 1991 (1963).
8. M. J. SIENKO and T. B. N. TRUONG, *J. Am. Chem. Soc.* **83**, 3939 (1961).
9. M. J. SIENKO, Paper 21 in *Nonstoichiometric Compounds*, R. WARD Ed., Series 39, Am. Chem. Soc., Washington, D.C. (1963).
10. J. B. GOODENOUGH, The Oxy-Compounds of the Transition Elements in the Solid State, International Colloquium of the Centre de la Recherche, Bordeaux, France (1964).

11. A. FERRETTI, D. B. ROGERS and J. B. GOODENOUGH, *J. Phys. Chem. Solids*, **26**, (1965).
12. A. R. SWEEDLER, C. J. RAUB and B. T. MATTHIAS, *Phys. Letters* **15**, 108 (1956).
13. J. F. SCHOOLEY, W. R. HOSLER and M. L. COHEN, *Phys. Rev. Letters* **12**, 474 (1964).
14. A. R. SWEEDLER, private communication.
15. F. GALASSO and W. DARBY, *Inorg. Chem.* **1**, 71 (1965).
16. D. D. GLOWER and R. C. HECKMAN, *J. Chem. Phys.* **41**, 877 (1964).
17. E. K. WEISE and I. A. LESK, *J. Chem. Phys.* **21**, 801 (1953).
18. E. J. W. VERWEY, P. W. HAAIJMAN, F. C. ROMEYN and G. W. VAN OOSTERHOUT, *Philips Res. Rpts.* **5**, 173 (1950).
19. G. H. JONKER and J. H. VAN SANTEN, *Physica* **16**, 337 (1950).
20. G. H. JONKER, *Physica* **20**, 1118 (1954).
21. P. W. HAAIJMAN, R. W. DAM and H. A. KLASSENS, *Method of Producing Semiconducting Materials*, German Pat. 929,350, June 23 (1955).
22. H. A. SAUER and S. S. FLASCHEN, *Proc. Electronic Components Symp.*, 7th Washington D.C., May, 41 (1956).
23. G. G. HARMAN, *Phys. Rev.* **106**, 1358 (1957).
24. O. SABURI, *J. Phys. Soc. Japan* **14**, 1159 (1959).
25. C. SABURI, *J. Am. Ceram. Soc.* **44**, 54 (1961).
26. J. TENNERY and R. COOK, *J. Am. Ceram. Soc.* **44**, 187 (1961).
27. W. T. PERIA, W. R. BRATSCHUN and R. P. FENITY, *J. Am. Ceram. Soc.* **44**, 249 (1961).
28. W. HEYWANG, *Solid State Elec.* **3**, 51 (1961).
29. F. M. RYAN and E. C. SUBBARAO, *App. Phys. Letters* **1**, 69 (1962).
30. G. GOODMAN, *J. Am. Ceram. Soc.* **46**, 49 (1963).
31. H. A. SAUER and J. R. FISHER, *J. Am. Ceram. Soc.* **43**, 297 (1960).
32. P. K. GALLAGHER, F. SCHEY and F. DIMARCELLO, *J. Am. Ceram. Soc.* **46**, 359 (1963).
33. P. K. GALLAGHER and F. SCHEY, *J. Am. Ceram. Soc.* **46**, 567 (1963).
34. J. W. DAVISSON and J. PASTERNAK, U.S. Naval Res. Labs. Memorandum Rpt. No. 1037 (March 1960).
35. General Ceramics, *Oxide Thermoelectric Materials Final Rpt.* (2 Feb. 60–2 Feb. 61), Contract NOBS-78414.

CHAPTER 5

FERROELECTRICITY

DURING World War II extensive investigations were conducted on titanates by von Hippel *et al.* at MIT under wartime restrictions.[1] After the war, the work was released along with studies conducted in England,[2-4] Russia[5-7] and Japan.[8] As a result of these investigations, barium titanate was found to be a ferroelectric.[7] The term ferroelectric was used because these materials are analogous in some ways to ferromagnetic materials. For example, when an alternating potential is applied to a capacitor containing a ferroelectric material, the instantaneous relation between charge and potential, or polarization and field, produces a hysteresis loop on a cathode-ray oscilloscope (see Fig. 5.1). Ferromagnetic

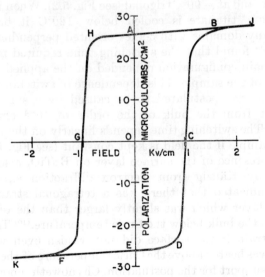

FIG. 5.1. Ferroelectric hysteresis loop.

materials also exhibit a hysteresis loop which represents the relation between magnetic induction B and magnetic field H.

A ferroelectric has been defined as a dielectric having a spontaneous polarization which can be reversed in sign. It therefore must have a polar structure with no center of symmetry. In changing the direction of the polar axis the structure must pass through an intermediate non-polar stage and the polar structure is a distortion of this more symmetrical form. The structure of ferroelectric materials becomes less distorted as the temperature increases and undistorted at and above a temperature called the Curie point.

5.1. Ternary Perovskites

Titanates

Barium titanate has the ideal cubic perovskite structure above the Curie point, but on cooling below this temperature the oxygen and titanium ions shift to form a tetragonal structure with the c-axis about 1% longer than the other two. At about 0°C the symmetry of the crystal becomes orthorhombic, and at $-90°C$ trigonal (see Fig. 5.2). When a crystal of barium titanate is cooled below 120°C it breaks up into many domains with c-axes oriented perpendicularly.

Merz[9] found that the switching time required to change the domain configuration depended on the applied field and the size of the sample. This dependence of switching time on thickness was postulated to be caused by a surface layer different from the bulk of the order of 10^{-4} cm on the crystal. The switching time depends linearly on the thickness of the sample if the field is kept constant (see Fig. 5.3).

The presence of the surface layer on $BaTiO_3$ also was reported by Känzig from electron diffraction experiments which indicated that there was a tetragonal strain in the surface layer which was slightly larger than the tetragonal strain in the bulk below the Curie temperature.[10] The tetragonal strain in the surface did not vanish even when the crystal was heated above the Curie temperature of the bulk. As further support for the postulation, Chynoweth reported that pyroelectric currents could be produced in single crystals of

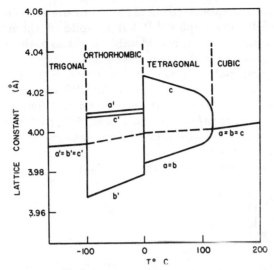

Fig. 5.2. Lattice constants of BaTiO$_3$ as functions of temperature (after H. F. Kay and P. Vousden, *Phil. Mag.* **40**, 229 (1948)).

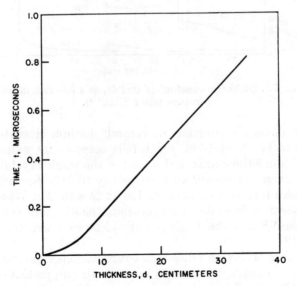

Fig. 5.3. Switching time for infinite field vs. thickness of the sample (after Merz[9]).

barium titanate above the Curie temperature even though no electric field was applied,[11] but in spite of these studies, there still appears to be considerable doubt as to the existence of the space-charge layer on the surface.

The tetragonal distortion in the structure of $BaTiO_3$ resulted in the formation of dipoles. Merz[12] measured the dipole moment of $BaTiO_3$ single crystals and found it to be 18×10^{-6} coulombs/cm^2 at 120°C and 26×10^{-6} at ambient temperature. In addition, he[13] also measured the dielectric constant and found that it was much greater perpendicular to the c-axis than along it (see Fig. 5.4).

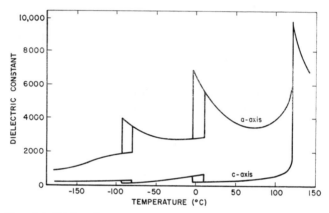

Fig. 5.4. Dielectric constant of $BaTiO_3$ as a function of temperature (after Merz[13]).

The dielectric constant of ceramic barium titanate was found to be about 1500, which falls between the values obtained along the c-axis and a-axis of the single crystal. The value decreases slowly with frequency to 10^6 c/s, and then decreases further to a value of 126 at 24×10^3 Mc. When the measurement is made under pressure, the dielectric constant increases[14] and the Curie point shifts to lower temperatures.[15]

Because of this high dielectric constant, experimental studies have been conducted to improve the properties of barium titanate for energy converter and capacitor applications.[16] A figure of merit for capacitor applications is denoted as (time constant) RC which is the time for a capacitor to discharge to $1/e$ of the charging voltage. Time constants were measured for various grades of $BaTiO_3$ and for $BaTiO_3$ with different additives. The results are presented in Table 5.1.

TABLE 5.1. *Measurements on* $BaTiO_3$ *Compositions Fired 1 hr at* $1450°C$ *(after Hoh and Pirigy*[16]*)*

$BaTiO_3$ starting material grade	Additive	RC in sec.
1. C.P.	none	10
2. Commercial capacitor	none	52
3. Commercial piezoelectric	none	220
4. High purity	none	700
5. Commercial capacitor	0.1 wt% Cr_2O_3	3,000
6. High purity	0.1 wt% Cr_2O_3	12,500
7. High purity	0.15 wt% Cr_2O_3	10,500
8. Commercial capacitor	3.0 wt% $CaSnO_3$	9,800
9. Commercial capacitor	0.1 wt% Cr_2O_3 +1 wt% $CaSnO_3$	12,000
10. High purity	0.1 wt% Cr_2O_3 as CrF_3	3,500
11. Commercial capacitor	0.5 mole % UO_2 as U_3O_8	10,500
12. Commercial capacitor	0.5 mole % UO_2 as UF_4	8,000
13. Commercial capacitor	0.5 mole % UO_2 as UF_4 + 3 mole % $CaSnO_3$	13,000

The results showed that the additives improved the time constant by several orders of magnitude. The compounds Cr_2O_3 and $CaSnO_3$ improved RC only at room temperature, but the solid fluoride additives showed the most promise.

MacChesney *et al.*[17] improved the stability of the properties with respect to temperature by adding La_2O_3. The temperature coefficient of capacitance and dissipation factor was reduced to low values by adding 1 mole % La_2O_3.

In other studies, increased stability of $BaTiO_3$ has been

reported for materials with the following additives: 1% $NiZrO_3$, $(Ti, Zr, Sn)O_2 + Bi_2O_3$, 6% $(Bi_2O_3.2ZrO_2)$, $(Zr, Sn)O_2 +$ $+ Li_2O$, $(Zr, Sn)O_2 + Al_2O_3$, $2.37(CaTiO_3) + 1.43(Bi_2SnO_5) +$ $+ 2.85SnO_2$, $1 - 25(CaZrO_3)$, and $41Fe_2O_3 + 10TiO_2 + 9FeO +$ $+ 5NiO + 5CoO$.

High dielectric constants have been produced in barium titanate after reduction and re-oxidation and with metal additives. However, it is probable that the high dielectric constants reported for these materials are high because of semiconductivity of the samples or the presence of a thin nonconducting skin on a conducting medium.

Khodakov[18] attempted to modify the properties of $BaTiO_3$ by using a fine particle size of $BaTiO_3$ $(1 - 20\mu)$ and found that the peaks in the dielectric constant versus temperature curves were flatter with decreasing particle size. This is an important method of obtaining materials whose dielectric constants are high and relatively independent of temperature.

Blinton and Havell[19] studied the properties of flame-sprayed barium titanate ceramic coatings. The coatings were predominantly the cubic phase. The metastable cubic phase transforms to the tetragonal phase by annealing the ceramic at 1400°C for 2 hr in air or helium. A study of the dielectric constant as a function of temperature revealed that the curves were much flatter than that for normal barium titanate.

Interesting dielectric materials also were prepared by crystallizing $BaTiO_3$ with feldspar $BaAl_2Si_2O_8$ by heat-treating glasses having compositions corresponding to $x\,BaTiO_3 +$ $+ (100 - x)\,(BaAl_2Si_2O_8)$. Herczog[20] proposed the mechanism of crystallization as being

$$\text{Glass } A \xrightarrow{\ 600° \text{ to } 800\ °C\ } BaTiO_3(Cr) + \text{glass } B$$

$$\text{Glass } B \xrightarrow{\ 750° \text{ to } 1000\ °C\ } BaAl_2Si_2O_8(Cr) + BaTiO_3(Cr)$$

If the heating rate and the final temperature of heat treatment were controlled the particle size could be varied from 0.01 to 1μ. The size of the particles were quite uniform for any treatment. For particles below 0.2μ the dielectric constant was nearly independent of temperature, and materials with a particle size of 1μ were found to have the highest dielectric constant. Resistivity and dielectric strength were high compared with ceramic materials.

Barium titanate also exhibits piezoelectric properties which means that electric polarization takes place when it is subjected to mechanical strain, and inversely the material mechanically deforms upon application of an electric field. This effect reverses in sign upon reversal of the electric field. This is in contrast to electrostriction exhibited by all dielectrics under an applied field.

Determination of the piezoelectric properties of single-crystal $BaTiO_3$ gave 950×10^{-8} statcoul/dyne for d_{33} and -310×10^{-8} statcoul/dyne for d_{31} where d_{33} is the proportionality between the charge developed on the two faces perpendicular to its c-axis and d_{31} is the proportionality between the charge on the same two faces and the force applied when the force is perpendicular to the c-axis.[21-24]

In order to obtain this effect in ceramic barium titanate, it must be first poled by d.c. voltages of $20-30$ kV/cm which line up approximately 10% of the c-axis of the crystallites. The piezoelectric moduli measured for ceramics depend on the effectiveness of the poling operation.

Lead titanate is also a ferroelectric with a tetragonal distortion of the perovskite structure. Shirane and Pepinsky[25] studied the structure with X-ray and neutron diffraction. The results indicated a shift of 0.30 Å for the Ti ion along the c-axis and 0.47 Å for the Pb ions which were much larger displacements than those found in barium titanate (see Fig. 5.5). At 490°C, $PbTiO_3$ changes from a tetragonal form to a cubic form (see Fig. 5.6). The energy absorption at this transformation temperature is 1150 cal/mol.[26]

The dielectric constant of $PbTiO_3$ is about 100 at room temperature and reaches a peak of about 1000 at 490°C.[27] Studies on single crystals indicate that the Curie point is at 495°C. The domain structures of these single crystals were reported to be essentially the same as that observed in $BaTiO_3$.

The piezoelectric coefficient d_{33} for $PbTiO_3$ with additives was found to be less than 30×10^{-12} coulombs/newton except for specimens containing 1 mole % CaF_2 which gave values as high as 130×10^{-12} coulombs/newton.[28]

Strontium titanate has the cubic perovskite structure, and is not ferroelectric at room temperature even though it has a dielectric constant of 200. There are conflicting reports on

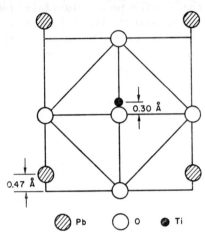

FIG. 5.5. Structure of PbTiO$_3$ (after G. Shirane, R. Pepinsky and G. B. Frazer, *Acta Cryst.* **9**, 131 (1956)).

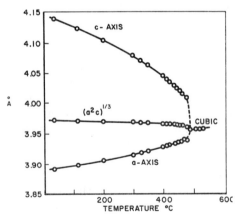

FIG. 5.6. Temperature dependence of volume and lattice constants for PbTiO$_3$ (after Shirane et al.[70]).

the possibility of it being ferroelectric at very low temperatures, even though Gränicher[29] did report the observation of a hysteresis loop with a field of 300 V/cm at 4°K. The spontaneous polarization was reported to be 3×10^{-6} coulomb/cm^2 and the remnant polarization 1×10^{-6} coulomb/cm^2.

Calcium titanate has an orthorhombic structure at room temperature, the structure becomes tetragonal at 600°C and cubic at 1000°C.[30] It has a room-temperature dielectric constant of 100, but it is not ferroelectric.

Niobates and Tantalates

Potassium niobate, $KNbO_3$, is the best-known compound of this group. The transition temperature on heating is $-12°C$ for the rhombohedral to orthorhombic transforma-

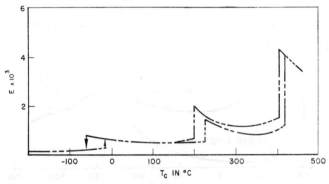

FIG. 5.7. Dielectric constant of $KNbO_3$ as a function of temperature (after Shirane et al.[31]).

tion, 224°C for the orthorhombic to tetragonal transformation, 412°C for the tetragonal to cubic transformation and $-59°$, 200° and 407°C on cooling (see Fig. 5.7). The loss tangent at these transitions is approximately 0.3.[31, 32] This is about ten times as high as the value for $BaTiO_3$. The values of saturation polarization have been obtained from hysteresis loops and found to be 0.9×10^{-6} coulombs/cm^2 at room temperature.

Sodium niobate is not ferroelectric and may be antiferroelectric, but ferroelectricity can be induced by the application of a strong field of the order of 10 kV/cm. Once ferroelectricity is induced, the crystals remain ferroelectric from $-55°C$ to 200°C. Vousden[33] and Francombe[34] indicated that there are several transformations in $NaNbO_3$ at 20°, 390°, 420°, 560° and 640°C. The last is the tetragonal to cubic transformation. The temperature dependence of the dielectric constant is given in Fig. 5.8.

A rectangular hysteresis loop with the same saturation polarization value of $BaTiO_3$ was investigated by Matthias[35] who reported the Curie temperature for $NaTaO_3$ to be 475°C.

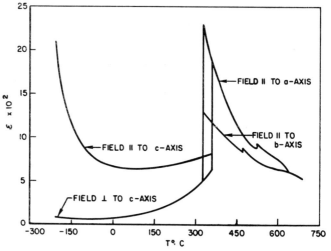

FIG. 5.8. Temperature dependence of dielectric constant of $NaNbO_3$ (after L. E. Cross and B. J. Nicholson[71]).

Potassium tantalate was reported as having a phase transformation between 10°K and 20°K and a hysteresis loop below 13°K[36]. Smolenskii reported a Curie temperature of 247°C for $RbTaO_3$.[37]

Zirconates and Hafnates

Lead zirconate, $PbZrO_3$, is antiferroelectric, that is although there is a dipole moment in each unit cell, the arrangement of the moments in adjacent cells is such as to cause a net dipole moment of zero. At 230°C, the symmetry of the structure changes from orthorhombic to cubic (see Fig. 5.9). The temperature variation of the dielectric constant of lead zirconate is given in Fig. 5.10.

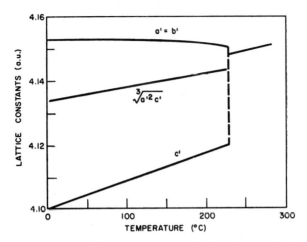

Fig. 5.9. Temperature variation of the lattice constants of lead zirconate (after E. Sawaguchi, *J. Phys. Soc. Japan* **8**, 615 (1953)).

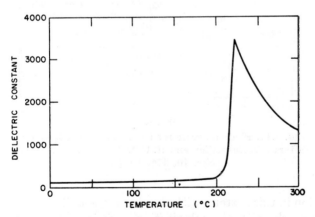

Fig. 5.10. Dielectric constant of lead zirconate as a function of temperature (after Roberts[55]).

Lead hafnate transforms to a new structure at 160°C and transforms at 210°C to a cubic form. The two lower forms are antiferroelectric and the high-temperature form is paraelectric.[38]

5.2. Solid Solutions

$BaTiO_3$–$SrTiO_3$

One of the most widely studied solid solution systems is that between $BaTiO_3$ and $SrTiO_3$. There is complete solid

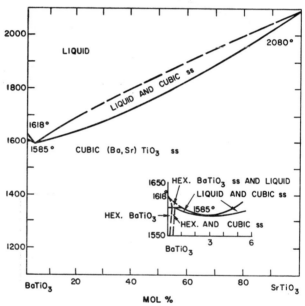

FIG. 5.11. Phase diagram for the $BaTiO_3$–$SrTiO_3$ system (after J. D. Basmajian and R. C. DeVries, *J. Am. Ceram. Soc.* **40**, 374 (1957)).

solution in this system, with the size of the unit cells decreasing linearly with the substitution of Sr in $BaTiO_3$[39, 40] (see Fig. 5.11). The Curie point also decreases with increasing amounts of strontium substitution (see Fig. 5.12). With the first additions of strontium the ambient temperature dielectric constant increases, reaches a maximum of 8000 at about 30 mole % addition, and then decreases. It is interesting that the low-temperature Curie point materials do not show the decrease in dielectric constant that is found for barium titanate at 10^9 cycles.[41-42] A minimum in the activation energy obtained from d.c. measurements was found at 40% $SrTiO_3$.[43]

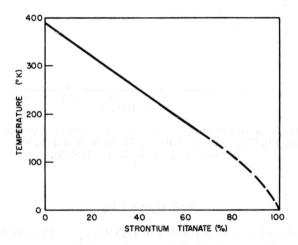

FIG. 5.12. Variation of Curie temperature as a function of composition, (Ba, Sr)TiO_3 (after Rushman et al.[39]).

$BaTiO_3$–$PbTiO_3$

Lead titanate also forms complete solid solution with barium titanate. The addition of lead titanate lowers the room temperature dielectric constant, increases the curie temperature (see Fig. 5.13),[34] decreases the d.c. resistivity and the activation energy[43], and improves the piezoelectric properties of barium titanate.[44, 45]

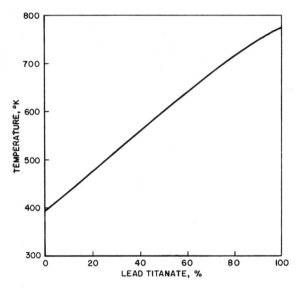

FIG. 5.13. Variation of Curie temperature with composition, (Ba, Pb)TiO$_3$ (after G. Shirane, S. Hoshino and K. Suzuki, *J. Phys. Soc. Japan* **5**, 456 (1950)).

BaTiO$_3$–CaTiO$_3$

Calcium titanate is soluble in BaTiO$_3$ up to 25 mole % and BaTiO$_3$ dissolves in CaTiO$_3$ to about the same extent (see Fig. 5.14). The addition of calcium to barium titanate lowers the room-temperature dielectric constant, and small amounts improve the piezoelectric properties.

BaTiO$_3$–(Ba, Sr, Ca)ZrO$_3$

The BaTiO$_3$–BaZrO$_3$ phase diagram is shown in Fig. 5.15. The addition of barium, strontium and calcium zirconate to barium titanate lowers the Curie point and broadens the maximum dielectric constant.[46]

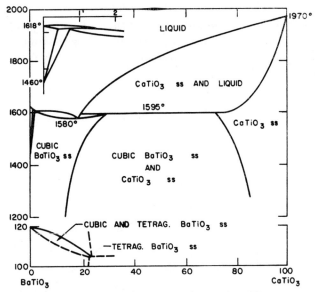

Fig. 5.14. Phase diagram for the $BaTiO_3$–$CaTiO_3$ system (after R. C. DeVries and R. Roy, *J. Am. Ceram. Soc.* **38**, 145 (1955)).

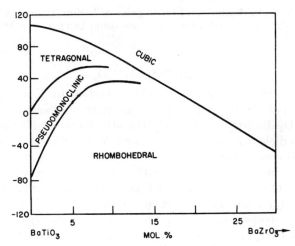

Fig. 5.15. Phase diagram for the system $BaTiO_3$–$BaZrO_3$ (after T. N. Verbitskaya, G. S. Zhdanov, Yu. N. Venevtsev and S. P. Solov'ev, *Kristallografiya*, **3**, 189 (1958)).

$BaTiO_3$–$BaSnO_3$

Dungan et al.[47] found that additions of $BaSnO_3$ to $BaTiO_3$ lowers the Curie temperature, and increases the unit cell size. Figure 5.16 shows the variation in Curie temperature with additions of $BaSnO_3$.

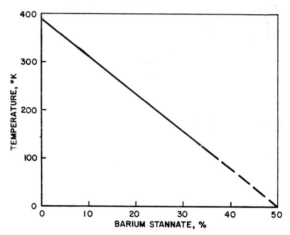

FIG. 5.16. Variation of Curie point of barium titanate with the addition of barium stannate (after Dungan et al.[47]).

$BaTiO_3$–$BaHfO_3$

A study of the $BaTiO_3$–$BaHfO_3$ system by Fresenko and Prokopolo[48] showed that it was quite similar to those of $BaTiO_3$–$BaZrO_3$ and $BaTiO_3$–$BaSnO_3$. Payne and Tennery[49] made dielectric measurements and X-ray diffraction studies in this system and found that the dielectric constant for each sample increased as the $BaHfO_3$ concentration was increased to 16 mole % $BaHfO_3$ and then decreased with further $BaHfO_3$ additions. They suggest that the ferroelectric–paraelectric transition for the composition containing 16 mole % $BaHfO_3$ was of second order and occurred between a ferroelectric rhombohedral phase and a paraelectric cubic phase.

Other $BaTiO_3$ Solid Solutions

More complex systems with barium titanate as one of the constituents have also been studied. The phase diagram for the system $BaSnO_3$–$BaTiO_3$$PbSnO_3$–$PbTiO_3$ is shown in Fig. 5.17.

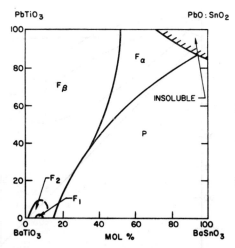

FIG. 5.17. Phase diagram for the $PbTiO_3$–$BaTiO_3$–$PbO:SnO_2$–$BaSnO_3$ system, P = paraelectric, cubic phase, F_α = ferroelectric, rhombohedral phase, F_β = ferroelectric, tetragonal phase, F_1 = ferroelectric, rhombohedral phase, F_2 = ferroelectric, orthorhombic phase (after T. Ikeda, *J. Phys. Soc. Japan* **14**, 1292 (1959)).

$PbTiO_3$–$KNbO_3$

The phase diagram of the $PbTiO_3$–$KNbO_3$ system and dielectric properties of the solid solutions were determined by Tien *et al.*[50] Where solid solution was formed between compounds with one common cation, the Curie temperature varied more or less linearly with composition. Figure 5.18 presents this data for the $KNbO_3$–$NaNbO_3$, $PbTiO_3$–$PbZrO_3$, $PbTiO_3$–$BaTiO_3$, $PbTiO_3$–$NaNbO_3$ and the $KNbO_3$–$KTaO_3$ systems. In the $PbTiO_3$–$KNbO_3$ system where no ions are common the preparation of homogeneous specimens proved to be difficult. Tien et al. feel that the minimum that exists is common for all systems which involve ferroelectric or antiferroelectric compounds not containing a common ion, point-

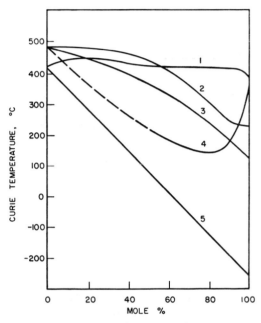

FIG. 5.18. Curie temperatures in perovskite systems: (1) $KNbO_3$–$NaNbO_3$, (2) $PbTiO_3$–$PbZrO_3$, (3) $PbTiO_3$–$BaTiO_3$, (4) $PbTiO_3$–$NaNbO_3$, (5) $KNbO_3$–$KTaO_3$ (after Tien et al.[50])

ing out that this is true of the $BaTiO_3$–$PbZrO_3$, $NaNbO_3$–$PbZrO_3$, $NaNbO_3$–$PbTiO_3$ as well as the $PbTiO_3$–$KNbO_3$ system.

$PbZrO_3$–$PbTiO_3$

Addition of barium, strontium and titanium ions destroys the antiferroelectric properties of $PbZrO_3$. Figure 5.19 shows a phase diagram of the $PbZrO_3$–$PbTiO_3$ system. The addition of lead titanate to lead zirconate appears to lower the dielectric constant and increases the temperature of its maximum.[51]

Jaffe et al.[52] found that in systems where compositional boundaries exist between ferroelectric phases of slightly differing structure the induced piezoelectric effects are enhanced as the composition approaches the phase boundary. In the system lead titanate–lead zirconate, lead titanate–lead oxide:tin oxide, lead zirconate–lead oxide:tin oxide and the lead titanate–lead hafnate, one of the compositions 45%

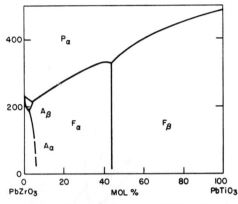

Fig. 5.19. Phase diagram for the $PbTiO_3$–$PbZrO_3$ system, P = paraelectric, cubic phase, A_α = antiferroelectric, orthorhombic phase, A_β = antiferroelectric, F_α = ferroelectric, rhombohedral phase, F_β = ferroelectric, tetragonal phase (after E. Sawaguchi, *J. Phys. Soc. Japan* **8**, 615 (1953)).

$PbTiO_3$–55% $PbZrO_3$ has a Curie temperature of 340°C and a radial coupling coefficient of 0.3 at 275°C which is twice that of barium titanate. Another containing 47.25% $PbTiO_3$, 22.75% $PbZrO_3$ and 30% $PbO:SnO_2$ had the highest piezoelectric coefficient (d_{31}) of all compositions studied—74 $\times 10^{-12}$ coulomb/newton.

Ikeda[53] reported that the addition of $LaFeO_3$ to $Pb(Ti, Zr)O_3$ ceramics near the phase boundary improved the piezoelectric performance. In more detailed studies Ikeda[54] showed that improved piezoelectric ceramics were obtained when $Pb(Ti,Zr)O_3$ was modified by the addition of $A^{1+}B^{5+}O_3$ (A = K, Na, B = Sb, Bi) or $A^{3+}B^{3+}O_3$ (A = Bi, La; B = Fe, Al,

Cr). Dielectric constants above 1500 and radial coupling coefficients above 0.6 were obtained with Na^+ and Sb^{5+} substitution.

$PbZrO_3$–$BaZrO_3$

When barium replaces lead in $PbZrO_3$, the dielectric maximum of lead zirconate is shifted to lower temperatures.[55] In addition, the maximum dielectric constant becomes great-

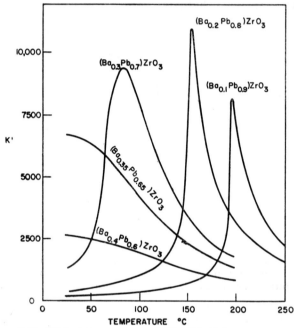

FIG. 5.20. Dielectric constant of (Ba, Pb)ZrO_3 compositions at varying temperatures (after Roberts[55]).

er and the high values are maintained over a broader range of temperature (see Fig. 5.20). The largest value is obtained for a composition of $(Ba_{0.20}Pb_{0.80})ZrO_3$ and when still larger percentages of barium are substituted, the dielectric maximum is lowered, but the temperature range of high dielectric constant values become broader. The dielectric constant of $Ba_{0.35}Pb_{0.65}ZrO_3$ stays above 6000 from room temperature up to 60°C making it useful for capacitor application.

5.3. COMPLEX PEROVSKITES

Viskov et al.[56] reported that the compounds $Ba(Bi_{0.5}Nb_{0.5})O_3$, $Ba(Bi_{0.5}Ta_{0.5})O_3$, $Ba(Bi_{0.5}U_{0.5})O_3$, $Ba(Bi_{0.67}W_{0.33})O_3$ and $Ba(Bi_{0.5}Mo_{0.5})O_3$ had distorted unit cells and high dielectric constants which peaked with temperature. They reported anomalies in dielectric constants at temperatures of 420°, 410°, 320°, 400° and 260°C for the compounds con-

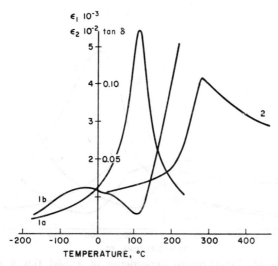

FIG. 5.21. Temperature dependence of ε for $Pb(Fe_{0.5}Nb_{0.5})O_3$ and $Pb(Y_{0.5}Nb_{0.5})O_3$: (1a and 1b) ε_1 and tan δ for $Pb(Fe_{0.5}Nb_{0.5})O_3$; (2) ε_2 for $Pb(Y_{0.5}Nb_{0.5})O_3$ (after Smolenskii et al.[57]).

taining Nb, Ta, V, W and Mo, respectively, and felt that the first three were ferroelectric and the last two antiferroelectric. However, evidence for this assumption was lacking.

Smolenskii et al.[57] found from studies on powder compacts that $Pb(Fe_{0.5}Nb_{0.5})O_3$ and $Pb(Yb_{0.5}Nb_{0.5})O_3$ might be ferroelectrics with Curie temperatures of 112°C and 280°C, respectively. The temperature dependence of permittivity and loss tangent for these compounds are shown in Fig. 5.21. The compound $Pb(Fe_{0.5}Nb_{0.5})O_3$ exhibited a hysteresis loop, but $Pb(Yb_{0.5}Nb_{0.5})O_3$ only showed a maximum in its permittivity and therefore may be antiferroelectric.

Smolenskii et al.[58] also found that $Pb(Sc_{0.5}Nb_{0.5})O_3$ and $Pb(Sc_{0.5}Ta_{0.5})O_3$ were ferroelectric materials. The dielectric constant of the niobium- and tantalum-containing compounds exhibited a maximum at 90° and 20°C, respectively (see Fig. 5.22). The characteristic drop in the loss tangent corresponds to a maximum of the dielectric constant. Hysteresis loops were obtained for both compounds at 18°C. The spontaneous polarization for $Pb(Sc_{0.5}Nb_{0.5})O_3$ at 18° equaled 3.6 microcoulombs and the coercive force 6 kV/cm.

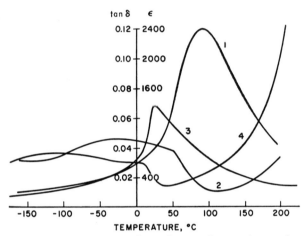

FIG. 5.22. Temperature dependence of ε and tan δ of $Pb(Sc_{0.5}Nb_{0.5})O_3$ and $Pb(Sc_{0.5}Ta_{0.5})O_3$: (1, 2) ε and tan δ of $Pb(Sc_{0.5}Nb_{0.5})O_3$; (3, 4) ε and tan δ of $Pb(Sc_{0.5}Ta_{0.5})O_3$ (after Smolenskii et al.[58]).

Maxima in dielectric constant with temperatures were observed for $Pb(Lu_{0.5}Nb_{0.5})O_3$, $Pb(Yb_{0.5}Ta_{0.5})O_3$ and $Pb(In_{0.5}Nb_{0.5})O_3$ at 280° for the first two compounds and 90° for the last, but they were probably antiferroelectric.[59] Studies on single crystals of $Pb(Co_{0.5}W_{0.5})O_3$[60] showed that it had a maximum in its dielectric constant at 32°C and exhibited a double hysteresis-loop characteristic of an antiferroelectric material.

Measurements on $Pb(Fe_{0.67}W_{0.33})O_3$ and $Pb(Fe_{0.5}Ta_{0.5})O_3$ indicate that these compounds are true ferroelectrics, while $Pb(Mg_{0.5}W_{0.5})O_3$ appears to be antiferroelectric.[61] A sharp permittivity peak was observed at 38°C in the dielectric constant–temperature curve for $Pb(Mg_{0.5}W_{0.5})O_3$. The compounds $Pb(Fe_{0.67}W_{0.33})O_3$ and $Pb(Fe_{0.5}Ta_{0.5})O_3$ exhibited hysteresis loops at liquid-oxygen temperature and the latter compound exhibited a maximum in its dielectric constant at $-30°C$.

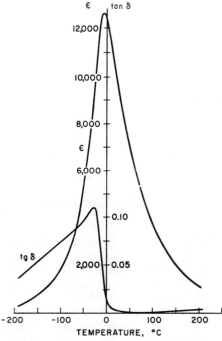

FIG. 5.23. Temperature dependence of ε and $\tan \delta$ values for $Pb(Mg_{0.33}Nb_{0.67})O_3$ (1 kc) (after Smolenskii et al.[63]).

Single crystals of $Pb(Ni_{0.33}Nb_{0.67})O_3$ were prepared by Myl'nikova and Bokov and reported to be ferroelectric. A maximum was observed in a plot of its dielectric constant versus temperature, and a hysteresis loop was observed at a temperature of -196 °C.[62]

Smolenskii and Agranovskaya[63] studied a large number of perovskite-type compounds and found two new ferroelectric materials, $Pb(Mg_{0.33}Nb_{0.67})O_3$ and $Pb(Ni_{0.33}Nb_{0.67})O_3$. Figure 5.23 shows the temperature dependence of the dielectric constant and loss tangent for $Pb(Mg_{0.33}Nb_{0.67})O_3$. The dielectric constant reached a maximum value of 12,600 at $-15°C$ when a frequency of 1 kc/s was used in the measurements. A hysteresis loop was observed at $-130°C$ and a spontaneous polarization value of 14×10^{-6} coulombs was calculated. The compound $Pb(Ni_{0.33}Nb_{0.67})O_3$ also appears to be a ferroelectric material.

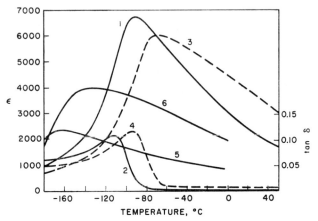

FIG. 5.24. The temperature dependences of ε and tan δ: (1, 2) ε and tan δ for single crystals $Pb(Mg_{0.33}Ta_{0.67})O_3$ (1 kc); (3, 4) ε and tan δ for single crystals $Pb(Co_{0.33}Nb_{0.67})O_3$ (1 kc); (5) ε for single-crystal $Pb(Ni_{0.33}Ta_{0.67})O_3$ (450 kc), (6) ε for single-crystal $Pb(Co_{0.33}Ta_{0.67})O_3$ (1 kc), (after Bokov et al.[64]).

Bokov and Myl'nikova[64] prepared single crystals of compounds $Pb(Ni_{0.33}Ta_{0.67})O_3$, $Pb(Mg_{0.33}Ta_{0.67})O_3$, $Pb(Co_{0.33}Nb_{0.67})O_3$, $Pb(Co_{0.33}Ta_{0.67})O_3$ and $Pb(Zn_{0.33}Nb_{0.67})O_3$ and showed that the compounds were ferroelectrics. The temperature dependence of the dielectric constant and loss tangent for these compounds are shown in Figs. 5.24 and 5.25. The loss tangent maximum occurred at a slightly lower temperature than the permittivity maximum which is characteristic of ferroelectrics. The authors attributed the differences of the phase transitions to the absence of ordering of the B ions in the octahedral positions. The data obtained from the hysteresis loops are the following for $Pb(Co_{0.33}Nb_{0.67})O_3$, $E_{max} = 28$ kV/cm, at $t = -150°C$, for $Pb(Zn_{0.33}Nb_{0.67})O_3$,

$E_{max} = 38$ kV/cm, at $t = 20°C$, for $Pb(Mg_{0.33}Ta_{0.67})O_3$, $E_{max} = 45$ kV/cm, at $t = 182°C$, for $Pb(Co_{0.33}Ta_{0.67})O_3$, $E_{max} = 70$ kV/cm, at $t = -196°C$, and for $Pb(Ni_{0.33}Ta_{0.67})O_3$, $E_{max} = 150$ kV/cm, at $t = -196°C$. The Curie temperatures of compounds $Pb(Mg_{0.33}Nb_{0.67})O_3$, $Pb(Mg_{0.33}Ta_{0.67})O_3$, $Pb(Co_{0.33}Nb_{0.67})O_3$, $Pb(Co_{0.33}Ta_{0.67})O_3$, $Pb(Ni_{0.33}Nb_{0.67})O_3$, $Pb(Ni_{0.33}Ta_{0.67})O_3$ and $Pb(Zn_{0.33}Nb_{0.67})O_3$ are $-12°C$, $-98°C$, $-70°C$, $-140°C$, $-120°C$, $-180°C$ and $140°C$, respectively.

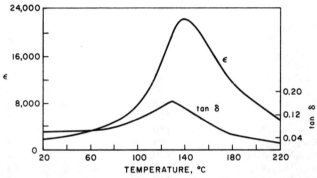

FIG. 5.25. The temperature dependence of ε and $\tan \delta$ of single crystal $Pb(Zn_{0.33}Nb_{0.67})O_3$ (1 kc) (after Bokov et al.[64]).

Johnson et al.[65] conducted an extensive study of complex ferroelectric solid solutions $Pb(Sc_{0.5}Nb_{0.5})_x \phi_{1-x}O_3$ where ϕ is either Ti, Zr, or Hf, finding that the maximum value of the spontaneous polarization of a system decreased as Ti was replaced by Zr and Zr by Hf and the Curie temperature increased as ϕ increased. In addition, the Curie points decreased in value as Ti was replaced by Zr and Zr by Hf. A complete list of ferroelectric compounds, Curie points and polarization values are given in Table 5.2.

5.4. Effect of Nuclear Irradiation

In recent years there has been considerable interest in the effect of nuclear radiation on the properties of ferroelectric materials. In one study, Glower and Hester[66] found that

TABLE 5.2. *Ferroelectrics Data*

	T °C	Ps (at T °C)	Refs.
$BaTiO_3$	120	10^{-6}	69
		26.0 (23)	
$PbTiO_3$	490	750.0 (23)	70
$KNbO_3$	435	0.9 (23)	35
		30 (250)	
$KTaO_3$	−260		35
$NaNbO_3$	−200		71
$CdTiO_3$	−218		72
$Pb(Cd_{0.5}W_{0.5})O_3$			73
$Pb(Sc_{0.5}Nb_{0.5})O_3$	90	3.6 (18)	58
$Pb(Sc_{0.5}Ta_{0.5})O_3$	26		58
$Pb(Fe_{0.67}W_{0.33})O_3$	−75		74
$Pb(Fe_{0.5}Nb_{0.5})O_3$	112		75
$Pb(Fe_{0.5}Ta_{0.5})O_3$	−30		76
$Pb(Mg_{0.33}Nb_{0.67})O_3$	−12	14 (−130)	64
$Pb(Ni_{0.33}Nb_{0.67})O_3$	−120		64
$Pb(Ni_{0.33}Ta_{0.67})O_3$	−180		64
$Pb(Mg_{0.33}Ta_{0.67})O_3$	−98		64
$Pb(Co_{0.33}Nb_{0.67})O_3$	−70		64
$Pb(Co_{0.33}Ta_{0.67})O_3$	−140		64
$Pb(Zn_{0.33}Nb_{0.67})O_3$	140		64

nuclear reactor irradiation of single crystal of $BaTiO_3$ produced increases in the coercive field (E_c) and a decrease in the remnant polarization (P_r). Crystals in the polarized state during irradiation were more resistant to radiation damage than were virgin crystals and the radiation damage rate was only slightly dependent upon crystal thickness. In their conclusions, they interpreted the changes in E_c and P_r in terms of a radiation model involving a build-up of a space charge due to the trapping of ionized carriers in the domain walls of the crystals.

Hilczer[67] reported that irradiation of barium titanate with 10^{19} neutrons/cm^2 reduced the dielectric constant by up to 40%. Doses of 10^{14}–10^{19} neutrons/cm^2 of (a) pile neutrons containing 10% fast neutrons; and (b) fast neutrons only that had passed through a 0.4-mm Cd foil gave the same effect, implying that fast neutrons produced the damage.

Schenk[68] found that tetragonal $BaTiO_3$ when irradiated with neutron doses of 4.2×10^{18} neutrons/cm^2 at 35°C transformed to cubic with radiation damage. The lattice expanded 2.26% for a and 1.17% for c.

5.5. Applications of Ferroelectric Materials

Use of Ferroelectric Properties

The high dielectric constants and ferroelectric behavior of perovskite-type compounds are probably the most important properties they exhibit.

Materials such as barium titanate cannot be used as capacitors in tuned circuits or filters where high-frequency stability is needed. However, they can be used as by-pass, blocking and smoothing capacitors which present a low-impedance path to an alternating current above a certain frequency.

The room-temperature dielectric constant of barium titanate can be raised by the addition of strontium titanate which lowers the Curie temperature. However, a flattening of the permittivity peak by using mixtures of alkali titanates and zirconates is more important. To date most of these ceramic capacitors have been used at low voltages as by-pass capacitors.

In addition, the hysteresis loop of ferroelectric single crystals makes them of potential use for information storage in electronic computors. Ferroelectrics also have been used as dielectric amplifiers. These are analogous to magnetic amplifiers, which require magnetic materials with narrow rectangular hysteresis loops.

The International Telephone and Telegraph Corporation recently reported a new use for ferroelectrics. This method for producing high-voltage a.c. or d.c. power was based on the fact that the dielectric constant of a ferroelectric is sensitive to temperature at the Curie point. The capacitor is held at the Curie point and then heated, lowering the dielectric constant. Since the charge cannot decrease because of a diode in the circuit, there must be a rise in capacitor voltage. This increased voltage also means an increase of electrical energy, thus there is a conversion of heat energy into electrical energy. This scheme has been proposed for use in a space vehicle which spins so it alternately faces toward and away from the sun.

Use of Piezoelectric Properties

One of the best known uses of piezoelectric materials is for the measurement of force or pressure. However, since there is a current leakage with time, they have best been used for measuring dynamic pressures, in blast gauges and accelerometers. Recently, the Spark Pump, the heart of which is two lead zirconate–lead titanate piezoelectric elements, was introduced by Clevite as an ignition source for gasoline engines. These elements were capable of achieving voltages of 20,000 volts when mechanical pressure was applied.

Another use of piezoelectrics is the phonograph pick-up which transforms the mechanical energy from the phonograph needle to an electrical signal. One of the methods of accomplishing this involves the use of two piezoelectric plates combined into a sandwich which is subjected to bending forces.

Piezoelectric transducers have also been used for sound transmission and reception, and ultrasonic cleaning devices.

Quartz has been used for electric frequency control. The piezoelectric coupling causes a reaction with an electric driving circuit which forces the circuit to oscillate at an exact frequency.

In a wave filter application, the impedance property of a crystal near a resonance point is used to allow passage of an electric signal which falls within a prescribed band of frequencies, while other frequencies are not passed.

The selection of materials for these applications depends on their piezoelectric constant. Some constants are listed in Fig. 5.26 and the units in Table 5.3 for perovskite-type ceramics. Note that the k, electromechanical coupling coefficient, values show the relationship between the mechanical energy stored and the electrical energy applied or the electrical energy stored and the mechanical energy applied. A high coupling coefficient, that is the ability to convert from one form of energy to another, is desirable in most of the applications.

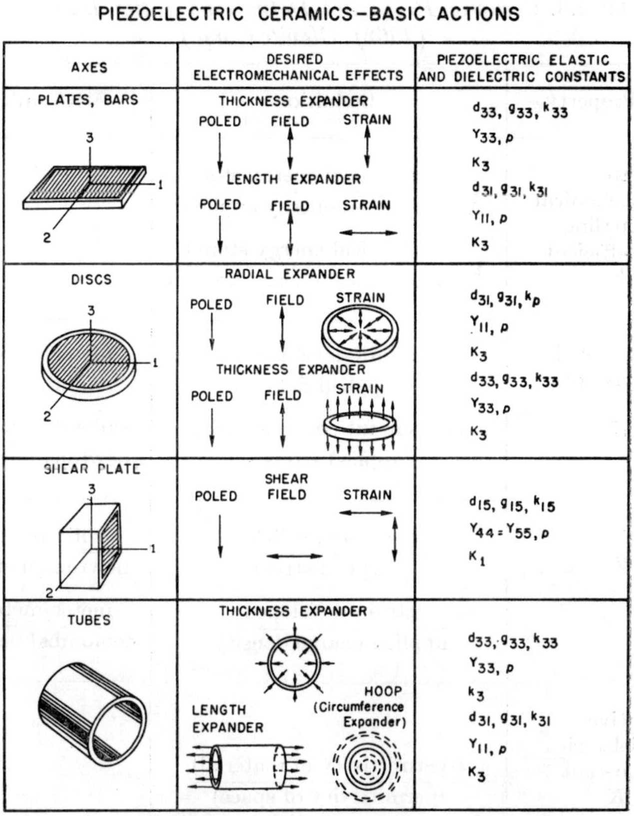

PIEZOELECTRIC CERAMICS—BASIC ACTIONS

AXES	DESIRED ELECTROMECHANICAL EFFECTS	PIEZOELECTRIC ELASTIC AND DIELECTRIC CONSTANTS
PLATES, BARS	THICKNESS EXPANDER (POLED, FIELD, STRAIN) / LENGTH EXPANDER (POLED, FIELD, STRAIN)	d_{33}, g_{33}, k_{33} $Y_{33,\rho}$ K_3 d_{31}, g_{31}, k_{31} $Y_{11,\rho}$ K_3
DISCS	RADIAL EXPANDER (POLED, FIELD, STRAIN) / THICKNESS EXPANDER (POLED, FIELD, STRAIN)	d_{31}, g_{31}, k_p $Y_{11,\rho}$ K_3 d_{33}, g_{33}, k_{33} $Y_{33,\rho}$ K_3
SHEAR PLATE	SHEAR (POLED, FIELD, STRAIN)	d_{15}, g_{15}, k_{15} $Y_{44} = Y_{55,\rho}$ K_1
TUBES	THICKNESS EXPANDER / LENGTH EXPANDER / HOOP (Circumference Expander)	d_{33}, g_{33}, k_{33} $Y_{33,\rho}$ k_3 d_{31}, g_{31}, k_{31} $Y_{11,\rho}$ K_3

FIG. 5.26. Piezoelectric ceramics—basic actions (after Bulletin 9247 (1962), Clevite Corp.).

The constant d is the ratio of the strain developed to the applied field or the short circuit charge density to the applied stress. The constant g is the ratio of the open circuit field to the applied stress or the strain developed to the applied charge density. A high d constant is desired for high amperage production and a high g constant for high voltage production. Other constants are K, relative dielectric constant in the material to space, and N, the frequency constant which is the controlling dimension times the resonant frequency.

TABLE 5.3. *Ceramic Properties Definitions (after Bulletin 9247 (1962), Clevite Corp.)*

Property	Definition	MKS units
Electromechanical coupling coefficient k	$\sqrt{\dfrac{\text{mechanical energy stored}}{\text{electrical energy applied}}}$ $\sqrt{\dfrac{\text{electrical energy stored}}{\text{mechanical energy applied}}}$	
Piezoelectric constants d	$\dfrac{\text{strain developed}}{\text{applied field}}$ $\dfrac{\text{short circuit charge density}}{\text{applied stress}}$	$\dfrac{\text{meter/meter}}{\text{volts/meter}}$ $\dfrac{\text{coulombs/meter}^2}{\text{newtons/meter}^2}$
g	$\dfrac{\text{open circuit field}}{\text{applied stress}}$ $\dfrac{\text{strain developed}}{\text{applied charge density}}$	$\dfrac{\text{volts/meter}}{\text{newtons/meter}^2}$ $\dfrac{\text{meter/meter}}{\text{coulombs/meter}^2}$
Relative dielectric constant K	$\dfrac{\varepsilon(\text{permittivity of material})}{(\text{permittivity of space})}$	
Modulus of elasticity Y	$\dfrac{\text{stress}}{\text{strain}}$	$\dfrac{\text{newtons/meter}^2}{\text{meter/meter}}$
Density ϱ		$\dfrac{\text{kg}}{\text{meters}^3}$
Frequency constant N	Controlling dimension $\times$ resonant frequency	cps-meters

The subscripts 1, 2 and 3 indicate the x-, y- and z-axes, respectively, and the subscripts 4, 5 and 6 represent a double subscript which stands for a plane. For example, 4 represents the yz-plane, 5 represents the xz-plane and 6 represents the xy-plane. The first subscript describes the direction of the field, the second the direction of the strain. For K the subscript p means planar coupling.

In the constant Y, the first subscript refers to the direction of stress and the second to the direction of the strain. The subscript for the dielectric constant refers only to the field. Table 5.4 presents the elastic, piezoelectric and dielectric properties of several ceramic compositions. The PZT materials are special commercial compositions of lead zirconate–lead titanate solid solutions. Using this data, the best materials for a particular application can be selected.

5.6. Theories of Ferroelectricity

Because of the importance of perovskite-type compounds as ferroelectrics, a brief review of theories which have been proposed to explain the phenomena associated with this property is presented. For more details, the original papers or the excellent treatment of this subject by Jona and Shirane in a book entitled *Ferroelectric Crystals*, published by Macmillan Company of New York, can be consulted.

Devonshire proposed a phenomenological theory for $BaTiO_3$.[76, 77] He assumes that $BaTiO_3$ in all forms can be considered to be a strained cubic crystal with a free energy which can be expressed as a function of temperature, stress and polarization. If the stress is initially taken as zero, an equation can be written as a series involving powers of the polarization P. The coefficients are functions of the stress-free condition and have subscripts X. Devonshire gives the equation with separate terms and coefficients for the components of polarization in the axial directions x, y, z. The equation is:

$$G_1 - G_{10} = \tfrac{1}{2}A^X(P_x^2+P_y^2+P_z^2) + \tfrac{1}{4}B^X(P_x^4+P_y^4+P_z^4) \\ + \tfrac{1}{6}C^X(P_x^6+P_y^6+P_z^6) + \tfrac{1}{2}D(P_y^2P_z^2+P_z^2P_x^2+P_x^2P_y^2),$$

TABLE 5.4. *Elastic, Piezoelectric and Dielectric Properties of Several Ceramic Compositions (after D. Berlincourt, B. Jaffe, H. Jaffe, and H. H. A. Krueger, IRE Nat'l Convention (1959))*

	95w % $BaTiO_3$ 5w % $CaTiO_3$	PZT-4	PZT-5
Coupling coefficients			
k_{33}	0.49	0.64	0.675
k_p	0.325	0.52	0.54
k_{31}	0.19	0.31	0.32
k_{15}	0.495	0.65	0.655
Piezoelectric constants			
d_{33}	150	256	320
d_{31}	−58	−111	−140
d_{15}	257	450	495
Free dielectric constants			
K_1	1280	1360	1285
K_3	1200	1200	1500
Frequency constants			
N_1	2290	1650	1500
N_3	2840	2000	1890
Elastic constants			
$1/s_{11}^E = Y_{11}^E$	11.6	8.15	6.75
$1/s_{33}^E = Y_{33}^E$	11.1	6.7	5.85
$1/s_{44}^E = Y_{44}^E$	4.4	2.6	2.0
Density	5.5	7.5	7.5
Mechanical Q	500	600	75
Curie point	115 °C	340°C	360 °C

where G_{10} is the value of G_1 for the unpolarized, unstressed crystal and the last term indicates that there is an interaction between the components of polarization along the x-, y- and z-axis.

From the equation given above, four sets of solutions which may correspond to minima in free energy can be obtained:

$$P_x = P_y = P_z = 0;$$
$$P_x = P_y = 0, \quad P_z \neq 0$$
$$P_z = 0, \quad P_x = P_y \neq 0$$
$$P_x = P_y = P_z \neq 0$$

These represent the cubic phase, in which the polarization is zero, and the tetragonal, orthorhombic (referred to monoclinic axes) and the rhombohedral forms respectively. The relative depths of the minima of the free energy function change with the coefficient A^X. If this decreases steadily and constant values are chosen for the coefficients, the temperatures at which the transitions occur are those actually observed.

On determining the constants, Devonshire drew theoretical curves for the spontaneous polarization, the free energy and the dielectric constants, over a range of temperatures. Qualitative agreement between calculated and experimental data was quite good.

Devonshire also gave equations for calculating spontaneous strain for a clamped crystal and discussed the effect of clamping on the nature of the transition. Probably one of the most important conclusions that can be made from the use of these equations is that the transition of the clamped crystal would be of the second order, even though that of the free crystal is of the first order.

Using Devonshire's approach of determining the coefficients for his equations from certain properties and employing them to predict others, a number of quantities were calculated. For example, the entropy change at the transition was determined for $BaTiO_3$ and $KNbO_3$, in reasonable agreement with experimental data.

Using a model approach, Mason and Matthias[78] suggest that the stable position for the Ti^{4+} ion in barium titanate is not at the center of the oxygen octahedra. Instead it is at any of the six positions which correspond to slight displacements from the center toward the oxygen ions. When the Ti^{4+} ion was in any of these positions the unit cell would have a dipole moment. However, if any dipole theory were correct, a number of polar liquids would be ferroelectrics which is not the case. In addition, with this theory it is not possible to obtain good agreement with experimental calculations.

Jaynes proposed a model in which oxygen ions are displaced rather than titanium ions[79] and also a theory which does not require the attribution of dipole moments to atomic displacement.[80] Only the electronic states of the TiO_6 octahedra are considered. The theory was adequate for determining the entropy change, but it predicts an infrared absorption line at $\sim 10\mu$, which was not detected.

Devonshire's model theory considers the dipole of an atom vibrating in the field of the neighbors. The dipole moment is not fixed in magnitude, but depends on the displacement from the equilibrium position.

Slater's theory[81] is similar to Devonshire's model theory. However, in addition, he assumes that each atom has an electronic polarization and titanium also has an ionic polarization. It predicts that the direction of spontaneous polarization is along the z-axis, but this is a disadvantage when it is applied to some structures.

In a structural approach, ferroelectricity and antiferroelectricity are associated with the off-center position of a high-valency cation in an octahedron. Megaw[82, 83] added the emphasis on the covalent bond character in the occurrence of ferroelectricity. The problem with Megaw's theory is that the origin of ferroelectricity is sought in abrupt changes in the character of the bonds at each transition. Like all of the theories described above, the structural approach has its limitation.

References

1. A. VON HIPPEL et al., RPTPB 4660 (1944).
2. W. JACKSON and W. REDDISH, Nature 156, 717 (1945).
3. H. MEGAW, Nature 155, 484 (1945).
4. H. P. ROOKSBY, Nature 155, 484 (1945).
5. B. M. VUL, Nature 156, 480 (1945).
6. B. M. VUL and I. M. GOLDMAN, C.R. Acad. Sci. URSS 46, 139 (1945).
7. V. GINZBURG, J. Phys. USSR 10, 107 (1946).
8. S. MIYAKE and R. UEDA, J. Phys. Soc. Japan 1, 32 (1945).
9. W. J. MERZ, J. Appl. Phys. 17, 938 (1956).
10. W. KÄNZIG, Phys. Rev. 98, 549 (1955).
11. A. G. CHYNOWETH, Phys. Rev. 102, 705 (1956).

12. W. J. Merz, *Phys. Rev.* **91**, 513 (1953).
13. W. J. Merz, *Phys. Rev.* **75**, 687 (1949).
14. B. M. Vul and L. Wereshchagin, *C.R. Acad. Sci. URSS* **48**, 634 (1945).
15. W. J. Merz, *Phys. Rev.* **78**, 52 (1950).
16. S. H. Hoh and F. E. Pirigy, *J. Am. Ceram. Soc.* **46**, 516 (1963).
17. J. B. MacChesney, P. K. Gallagher and F. V. DiMarcello, *J. Am. Ceram. Soc.* **46**, 197 (1963).
18. A. L. Khodakov, *Sov. Phys., Solid State* **2**, 1904 (1960).
19. J. L. Blinton and R. Havell, *Cer. Bull.*, Part II, **41**, 762 (1962).
20. A. Herczog, *J. Am. Ceram. Soc.* **47**, 107 (1964).
21. W. L. Cherry and R. Adler, *Phys. Rev.* **72**, 981 (1948).
22. W. P. Mason, *Phys. Rev.* **74**, 1134 (1948).
23. F. Sawaguchi and T. Akioka, *J. Phys. Soc. Japan* **4**, 117 (1949).
24. S. Nomura and S. Sawada, *J. Phys. Soc. Japan* **5**, 227 (1950).
25. G. Shirane and R. Pepinsky, *Phys. Rev.* **97**, 1179 (1955).
26. G. Shirane and E. Sawaguchi, *Phys. Rev.* **81**, 458 (1951).
27. G. Shirane and S. Hoshino, *J. Phys. Soc. Japan* **6**, 265 (1951).
28. T. Y. Tien and W. G. Carlson, *J. Am. Ceram. Soc.* **45**, 567 (1962).
29. H. Granicher, *Helv. Phys. Acta* **29**, 210 (1956).
30. H. F. Kay, *Rpt. Brit. Elec. Res. Assoc.*, Ref. L/T, 257 (1951).
31. G. Shirane, H. Danner, A. Pavlovic and R. Pepinsky, *Phys. Rev.* **93**, 612 (1954).
32. S. Triebwasser, *Phys. Rev.* **101**, 998 (1956).
33. P. Vousden, *Acta Cryst.* **4**, 545 (1951).
34. M. H. Francombe, *Acta Cryst.* **9**, 256 (1954).
35. B. T. Matthias, *Phys. Rev.* **75**, 1771 (1949).
36. J. K. Hulm, B. T. Matthias and A. Long, *Phys. Rev.* **79**, 885 (1950).
37. G. A. Smolenskii et al., *Dokl. Akad. Nauk SSSR*, **76**, 519 (1951).
38. G. Shirane and R. Pepinsky, *Phys. Rev.* **91**, 812 (1953).
39. D. F. Rushman and M. A. Strivens, *Trans. Faraday Soc.* **42A**, 231 (1946).
40. G. Durst, M. Grotenhurs and A. G. Barkow, *J. Am. Ceram. Soc.* **33**, 133 (1950).
41. J. G. Powles, *Nature* **162**, 655 (1958).
42. H. Iwayanagi, *J. Phys. Soc. Japan* **8**, 525 (1953).
43. S. Nomuro and S. Sawada, *J. Phys. Soc. Japan* **5**, 227 (1950).
44. E. J. Brajer, H. Jaffee and F. Kulcsár, *J. Acoust. Soc. Am.* **24**, 117 (1952).
45. D. A. Berlincourt and F. Kulcsár, *J. Acoust. Soc. Am.* **24**, 709 (1952).
46. R. G. Graf, *Ceram. Age* **58**, 16 (1951).
47. R. H. Dungan, D. F. Kane and L. R. Bickford, *J. Am. Ceram. Soc.* **35**, 318 (1952).
48. E. G. Fesenko and O. I. Prokopolo, *Sov. Phys., Cryst.* **6**, 373 (1961).
49. W. H. Payne and V. J. Tennery, *J. Am. Ceram. Soc.* **48**, 413 (1965).

50. T. Y. TIEN, E. C. SUBBARAO and H. HRIZO, *J. Am. Ceram. Soc.* **45**, 572 (1962).
51. G. SHIRANE, *Phys. Rev.* **86**, 219 (1952).
52. B. JAFFE, R. S. ROTH and S. MORZULLO, *J. Appl. Phys.* **25**, 809 (1954).
53. T. IKEDA, *J. Appl. Phys. Japan* **3**, 493 (1964).
54. T. IKEDA and T. OKANO, *J. Appl. Phys. Japan* **3**, 63 (1964).
55. S. ROBERTS, *J. Am. Ceram. Soc.* **33**, 66 (1950).
56. A. S. VISKOV, YU. N. VENEVTSEV and G. S. ZHDANOV, *Sov. Phys.-Doklady* **10**, 391 (1965).
57. G. A. SMOLENSKII, A. I. AGRANOVSKAYA, S. N. POPOV and V. A. ISUPOV, *Sov. Phys. Tech. Phys.* **3**, 1981 (1958).
58. G. A. SMOLENSKII, V. A. ISUPOV and A. I. AGRANOVSKAYA, *Sov. Phys., Solid State* **1**, 150 (1959).
59. M. F. KUPRIYANOV and E. G. FESENKO, *Sov. Phys., Cryst.* **10**, 189 (1965).
60. V. A. BOKOV, S. A. KIZHAEW, I. E. MYL'NIKOVA and A. G. TUTOV, *Sov. Phys., Solid State* **6**, 2419 (1965).
61. G. A. SMOLENSKII, A. I. AGRANOVSKAYA and V. A. ISUPOV, *Sov. Phys., Solid State* **1**, 907 (1959).
62. I. E. MYL'NIKOVA and V. A. BOKOV, *Sov. Phys., Cryst.* **4**, 408 (1960).
63. G. A. SMOLENSKII and A. I. AGRANOVSKAYA, *Sov. Phys., Solid State* **1**, 1429 (1960).
64. V. A. BOKOV and I. E. MYL'NIKOVA, *Sov. Phys., Solid State* **2**, 2428 (1961).
65. V. J. JOHNSON, M. W. VALENTA, J. E. DOUGHERTY, R. M. DOUGLASS and J. W. MEADOWS, *J. Phys. Chem. Soc.* **24**, 85 (1963).
66. D. D. GLOWER and D. L. HESTER, *J. Appl. Phys.* **36**, 2175 (1965).
67. B. HILCZER, *Phys. Status Solidi* **5**, 113 (1964).
68. M. SCHENK, *Phys. Status Solidi* **4**, 25 (1964).
69. A. VON HIPPEL, R. G. BRECKENRIDGE, F. C. CHESLEY and L. TISZA, *Ind. Eng. Chem.* **38**, 1097 (1946).
70. G. SHIRANE, S. HOSHINO and K. SUZUKI, *Phys. Rev.* **80**, 1105 (1950).
71. H. L. E. CROSS and B. J. NICHOLSON, *Phil. Mag.*, Ser. 7, **46**, 453 (1955).
72. G. A. SMOLENSKII, *Dokl. Akad. Nauk SSSR* **70**, 405 (1950).
73. V. A. ISUPOV and L. T. ELEM'YANOVA, *Kristallografiya* **11**, 776 (1966).
74. A. I. AGRANOVSKAYA, *Bull. Acad. Sciences, U.S.S.R., Phys. Sciences* **24**, 1271 (1960).
75. G. A. SMOLENSKII, A. I. AGRANOVSKAYA, S. N. POPOV and V. A. ISUPOV, *Sov. Phys. Tech. Phys.* **3**, 1981 (1960).
76. A. F. DEVONSHIRE, *Phil. Mag.* **40**, 1040 (1949).
77. A. F. DEVONSHIRE, *Phil. Mag. Suppl.* **3**, 85 (1954).
78. W. P. MASON and B. T. MATTHIAS, *Phys. Rev.* **74**, 1622 (1948).
79. E. T. JAYNES, *Phys. Rev.* **79**, 1008 (1950).
80. E. T. JAYNES, *Ferroelectricity*, Princeton University Press, (1953).
81. J. C. SLATER, *Phys. Rev.* **78**, 748 (1950).
82. H. D. MEGAW, *Acta Cryst.* **5**, 739 (1952).
83. H. D. MEGAW, *Acta Cryst.* **7**, 187 (1954).

CHAPTER 6

PHASE TRANSITIONS

THE phase transitions in perovskite-type compounds are often associated with a change in ferroelectric properties. Some of these were described in the previous chapter. In this chapter, an attempt will be made to describe phase transitions reported for perovskite-type compounds, whether or not they involve a ferroelectric transition. These phase transitions can be divided into those of the first kind which are associated with the absorption or liberation of heat along with discontinuous changes in entropy and in lattice parameters, and those of the second kind which involve a peak in heat capacity, in the coefficient of thermal expansion and in compressibility. A list of perovskite-type compounds with transition temperatures and references is given in Table 6.1.

6.1. TERNARY PEROVSKITES

The phase transitions in barium titanate are probably the best characterized. The cubic phase is stable down to 120°C, and below this temperature the tetragonal ferroelectric phase appears, and remains stable to about 5°C. Below 5°C a new phase is formed, which has a unit cell with orthorhombic symmetry and still is ferroelectric with the direction of spontaneous polarization being parallel to one of the original cubic $\langle 110 \rangle$ directions. At $-90°C$ another transition occurs and the symmetry of the structure becomes rhombohedral. The polar axis lies along one of the original cubic $\langle 111 \rangle$ directions. The thermal expansion of the cell parameters in each of these phases is shown in Fig. 5.2. The volume changes at the phase transitions with rising temperature are 0.0006,

TABLE 6.1. *Phase Transitions*

Compound	Transition temp., °C (to cubic at highest temp. listed)	References
$AgNbO_3$	325, 550	1
$AgTaO_3$	485	1
$BaTiO_3$	−90, 5, 120	2
$KNbO_3$	−10, 225, 435	3
$KTaO_3$	−260	4
$NaNbO_3$	−200, 354, 562, 640	5, 6
$NaTaO_3$	470	4, 7
$PbHfO_3$	163, 215	8
$PbTiO_3$	490	9
$PbZrO_3$	230	10, 11, 7
$SrTiO_3$	−220	7, 12, 13
$Pb(Co_{0.5}W_{0.5})O_3$	32	14, 15
$Pb(Cd_{0.5}W_{0.5})O_3$	410	16
$Pb(Lu_{0.5}Nb_{0.5})O_3$	280	16, 17
$Pb(Lu_{0.5}Ta_{0.5})O_3$	280	16, 17
$Pb(Yb_{0.5}Nb_{0.5})O_3$	280	17, 14, 18
$Sr(Co_{0.5}Mo_{0.5})O_3$	320	19
$Sr(Co_{0.5}W_{0.5})O_3$	400	19
$Sr(Fe_{0.5}Nb_{0.5})O_3$	250	20
$Sr(Fe_{0.5}Ta_{0.5})O_3$	250	20
$Sr(Ni_{0.5}Mo_{0.5})O_3$	230	19
$Sr(Ni_{0.5}W_{0.5})O_3$	300	19
$Sr(Y_{0.5}Nb_{0.5})O_3$	630	21
$Sr(Y_{0.5}Ta_{0.5})O_3$	640	21
$Sr(Zn_{0.5}Mo_{0.5})O_3$	420	19
$Sr(Zn_{0.5}W_{0.5})O_3$	430	19

0.014 and 0.062 Å^3 and the transition heats are 8, 22 and 50 cal/mole.

The cubic to tetragonal transition in barium titanate also is characterized by the appearance of domain patterns. Observation of ferroelectric domains between crossed nicols is an excellent method of studying phase transitions in ferroelectric materials. Examining crystals after etching and after electrostatically charged particles have been deposited on the surface are other methods which have been used to investigate the domain pattern.

Studies on tetragonal barium titanate using a polarizing microscope have shown that it is possible for domains to be polarized at 90° to each other. When the polar axis is perpendicular to the plane of a (001) plate, the domain is called a "c" domain, and when it lies within the plane of the plate, the domain is called an "a" domain. These domains are easily observed under polarized light, and their appearance or disappearance indicates the cubic to tetragonal or tetragonal to cubic transition.

If domains are polarized antiparallel to each other, they are called 180° domains and a field must be applied perpendicular to the polar axis to make them visible.

Considering phase transitions involving other $A^{2+}B^{4+}O_3$-type compounds, the structure of strontium titanate becomes cubic at $-220°C$ and that of calcium titanate changes to cubic at 1260°C. Lead titanate has a transition from a tetragonal ferroelectric phase to a cubic phase at 490°C, Fig. 5.6, the pseudotetragonal antiferroelectric $PbZrO_3$ phase undergoes a first-order phase transition to cubic at 230°C (Fig. 5.9) with a heat of transition of 440 cal/mole and $PbHfO_3$ shows transitions at 163°C and 215°C. The transition for $PbHfO_3$ at 215°C is probably an antiferroelectric transformation to cubic symmetry.

The phase transitions at 435°C, 225°C and $-10°C$ in $KNbO_3$ are quite similar to those found for $BaTiO_3$ except that the spontaneous strain in $KNbO_3$ is larger in all three phases. The transition energies are 190, 85 and 32 cal/mole, see Fig. 5.7, and the c/a for the unit cell of tetragonal $KNbO_3$ is 1.017 compared with the value of 1.010 for that of $BaTiO_3$. All of the phase changes are first order and exhibit temperature hysteresis. The structure of $KTaO_3$ changes to cubic symmetry at $-260°C$.

The structure of $NaNbO_3$ is monoclinic below $-200°C$ and is ferroelectric. Above this temperature the structure of $NaNbO_3$ has orthorhombic symmetry. This antiferroelectric phase changes to a nonpolar pseudotetragonal phase at

354°C. At 562°C the structure of $NaNbO_3$ becomes tetragonal, and cubic at 640°C. At room temperature, a polar structure can be induced by the application of a field and a double hysteresis loop can be observed.

A transition at 480° to cubic symmetry was found for $NaTaO_3$; however, no anomaly in the dielectric constant has been detected.

X-ray studies on $AgNbO_3$ indicate that the structure transforms from orthorhombic symmetry to tetragonal symmetry at 325°C and from tetragonal to cubic symmetry at 550°C. $AgTaO_3$ shows similar phase transitions at 370°C and 485°C.

6.2. Complex Perovskite-type Compounds

The onset of ferroelectricity in the complex perovskite-type compounds listed in Table 5.2 must be associated with phase transitions. However, very few such transitions have

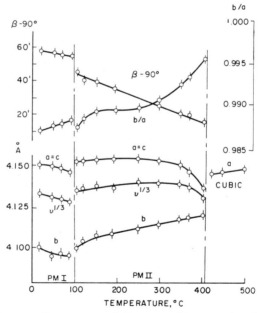

Fig. 6.1. Parameters of the perovskite pseudocell of $Pb(Cd_{0.5}W_{0.5})O_3$ as functions of temperature (after Roginskaya et al.[16]).

been reported with the ferroelectric data. In some cases, the back reflections in the X-ray patterns have been too poor to prove that small distortions existed in the ferroelectric phases.

Phase transitions for a few complex perovskite compounds have been studied. A pseudomonoclinic form of $Pb(Cd_{0.5}W_{0.5})O_3$ has been found to exist up to 100°C, at which temperature it transforms to a second monoclinic form and at 410°C it becomes cubic (see Fig. 6.1). The cubic form has the ordered

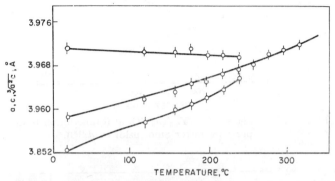

Fig. 6.2. Parameters of $Sr(Fe_{0.5}Nb_{0.5})O_3$ as a function of temperature (after Kupriyanov et al.[17]).

perovskite structure with a unit cell of "a" approximately equal to 8 Å. The authors feel that the monoclinic form of $Pb(Cd_{0.5}W_{0.5})O_3$ below 100°C is antiferroelectric.

The phase transformation of $Sr(Fe_{0.5}Nb_{0.5})O_3$ involves a change from a tetragonal to cubic form at 220°C (see Fig. 6.2). From the unit cell size given there does not appear to be any ordering of the B ions in this compound. A similar phase transition was reported for $Sr(Fe_{0.5}Ta_{0.5})O_3$.

The compound $Sr(Y_{0.5}Ta_{0.5})O_3$ was found to transform from a rhombohedral form to a cubic form at 640°C, similar to that in $Sr(Y_{0.5}Nb_{0.5})O_3$ (see Fig. 6.3).

Tetragonal to cubic phase transformations have been reported for $Sr(Ni_{0.5}W_{0.5})O_3$, $Sr(Co_{0.5}W_{0.5})O_3$, $Sr(Zn_{0.5}W_{0.5})O_3$, $Sr(Ni_{0.5}Mo_{0.5})O_3$, $Sr(Co_{0.5}Mo_{0.5})O_3$ and $Sr(Zn_{0.5}Mo_{0.5})O_3$ at temperatures of 300°, 400°, 430°, 230°, 320° and 420°C, respectively. The X-ray diffraction data indicates ordering of the B ions, but in Fig. 6.4 the cell size used for convenience was that of the simple perovskite structure. This figure shows the variation in cell size with temperature.

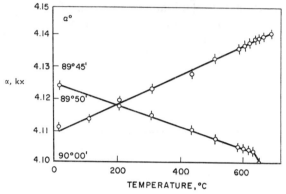

FIG. 6.3. Parameters of $Sr(Y_{0.5}Ta_{0.5})O_3$ as a function of temperature (after Smolenskii et al.[18]).

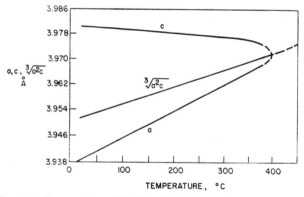

FIG. 6.4. Parameters of $Sr(Co_{0.5}W_{0.5})O_3$ as a function of temperature (after Kupriyanov et al.[19]).

Filip'ev and Fesenko studied the phase transition in $Pb(Co_{0.5}W_{0.5})O_3$ and found that it involved a change in symmetry from orthorhombic to cubic at 25°C. Superstructure was observed in both phases. A peak in the dielectric constant was seen at the phase transition, indicating that the compound was ferro- or antiferroelectric.

Antiferroelectric phase transitions from monoclinic to cubic symmetry were reported for the compounds $Pb(Lu_{0.5}Nb_{0.5})O_3$, $Pb(Lu_{0.5}Ta_{0.5})O_3$ and $Pb(Yb_{0.5}Nb_{0.5})O_3$ at 280°C. However, the transition temperature does not coincide with the temperatures at which the peak in the dielectric constants occurred for these compounds.

REFERENCES

1. M. H. FRANCOMBE and B. LEWIS, *Acta Cryst.* **11**, 175 (1958).
2. H. F. KAY and P. VOUSDEN, *Phil. Mag.* **40**, 1019 (1949).
3. G. SHIRANE, H. DANNER, A. PAVLOVK and R. PEPINSKY, *Phys. Rev.* **93**, 672 (1954).
4. B. T. MATTHIAS, *Phys. Rev.* **75**, 1771 (1949).
5. G. SHIRANE, R. NEWNHAM and R. PEPINSKY, *Phys. Rev.* **96**, 581 (1954).
6. M. H. FRANCOMBE, *Acta Cryst.* **9**, 256 (1956).
7. G. A. SMOLENSKII, *Zhur. Tekh. Fiz.* **20**, 137 (1950).
8. G. SHIRANE and R. PEPINSKY, *Phys. Rev.* **91**, 812 (1953).
9. G. SHIRANE and S. HOSKINO, *J. Phys. Soc. Japan* **6**, 265 (1951).
10. E. SAWAGUCHI, G. SHIRANE and Y. TAKAGI, *J. Phys. Soc. Japan* **6**, 333 (1951).
11. G. SHIRANE, E. SAWAGUCHI and Y. TAKAGI, *Phys. Rev.* **84**, 476 (1951).
12. G. A. SMOLENSKII, *Dokl. Akad. Nauk SSSR* **70**, 405 (1950).
13. G. A. SMOLENSKII, *Izvest. Akad. Nauk SSSR* Ser. Fiz. **20**, 166 (1956).
14. Y. TOMASHPOL'SKII and Y. N. VENEVTSEV, *Sov. Phys., Solid State* **6**, 2388 (1965).
15. V. A. BOKOV, S. A. KIZHAEV, I. E. MYL'NIKOVA and A. G. TUTOV, *Sov. Phys., Solid State* **6**, 2419 (1965).
16. Y. E. ROGINSKAYA and Y. N. VENEVTSEV, *Sov. Phys. Cryst.* **10**, 275 (1965).
17. M. F. KUPRIYANOV and E. G. FESENKO, *Sov. Phys. Cryst.* **10**, 189 (1965).
18. G. A. SMOLENSKII, A. I. AGRANOVSKAYA, S. W. POPOV and V. A. ISUPOV, *Sov. Phys. Techn. Phys.* **3**, 1981 (1958).
19. M. F. KUPRIYANOV and E. G. FESENKO, *Sov. Phys., Cryst.* **7**, 358 (1962).
20. M. F. KUPRIYANOV and E. G. FESENKO, *Sov. Phys., Cryst.* **6**, 639 (1961-2).
21. M. F. KUPRIYANOV and E. G. FESENKO, *Sov. Phys., Cryst.* **7**, 282 (1962).

CHAPTER 7

FERROMAGNETISM

THE common exchange energy in magnetic oxides is of the indirect (super-exchange) type. The energy between spins of neighboring metal ions in perovskite-type structures is often found to be negative so that antiparallel alignment has the lowest energy. It has been proposed that this alignment is caused by mutual interaction of the metal ions with the oxygen ion which is situated between them.

The magnetic oxides with the perovskite structure,

$La^{3+}Mn^{3+}O_3 - Ba^{2+}Mn^{4+}O_3$, $La^{3+}Mn^{3+}O_3 - Sr^{2+}Mn^{4+}O_3$, $La^{3+}Mn^{3+}O_3 - Ca^{2+}Mn^{4+}O_3$ and $La^{3+}Co^{3+}O_3 - Sr^{2+}Co^{4+}O_3$,

studied by Jonker and van Santen appear to be exceptions.[1] These studies indicated that a weak magnetic interaction between Mn^{3+} ions, a negative interaction between Mn^{4+} ions and strong positive interaction between Mn^{3+} and Mn^{4+} ions existed in the manganites. It was found that in the solid solution range $LaMn^{2+}O_3 - 25-35\%$ $AMn^{4+}O_3$ (A = Ca, Sr, or Ba) the magnetic saturation values agreed with the sum of the moments of Mn^{3+} and Mn^{4+} ions and the highest values of the Curie temperatures occurred in this region. The saturation magnetization at 90°K is given in Fig. 7.1 for mixed crystals of (La, Ca)MnO$_3$. A corresponding situation was found for compounds containing Co^{3+} and Co^{4+} ions, but not for compounds containing Cr^{3+} and Cr^{4+}, or with Fe^{3+} and Fe^{4+}, as the B ions were found to be antiferromagnetic[2]. All of these phases with the B metal ions in two valence states were highly conducting.

In order to avoid the high conductivity, Jonker studied $La(Mn^{3+}_{1-x}Cr^{3+}_x)O_3$ solid solutions which contained B ions of different elements with the same electronic configuration as the Mn^{3+} and Mn^{4+} ions. Positive magnetic interactions were found between $Mn^{3+}-Mn^{3+}$ and $Mn^{3+}-Cr^{3+}$ pairs at low temperatures. The saturation magnetization increased up to a composition of 30% $LaCrO_3$ and then the increasing number of $Cr^{3+}-Cr^{3+}$ strong negative interactions lowered the saturation magnetization. The maximum was found in the region of the transition point from a monoclinic to a pseudocubic structure.[3]

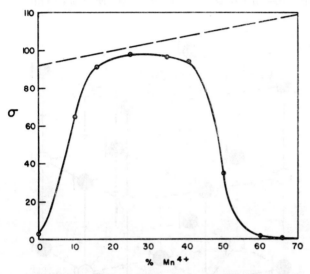

Fig. 7.1. Saturation magnetization at 20.4°K of $LaMnO_3$–$CaMnO_3$ (after G. H. Jonker, *Physica* 22, 707 (1956)).

In a similar study of the $(La,Ba)(Mn^{3+},Ti^{4+})O_3$ system, Jonker found that a maximum in the saturation magnetization existed at the composition which produced a change from a monoclinic to a cubic structure. Since the titanum ion is diamagnetic the positive interaction between Mn^{3+} ions is the only one possible in these phases.

Goodenough *et al.*[4, 5] investigated the $La(Mn_{1-x}Co_x)O_3$ series of phases and reported that the ferromagnetic saturation varied nearly linearly between $x = 0.20$ and $x = 0.70$. This was attributed to a positive interaction between Mn ions, with Co ions being in the diamagnetic low spin state. However, the similarity between the ferromagnetic saturation versus composition curve for the $La(Mn_{1-x}^{3+}Cr_x^{3+})O_3$ and $La(Mn_{1-x}Co_x)O_3$ series led Jonker to believe that there might be another explanation for the magnetic behavior of the former series.[6] His studies indicated that there was a rather strong positive interaction between the Co and Mn ions as well as the positive interaction between Mn^{3+} ions suggested by Goodenough. In addition, the results suggested that neighboring pairs of Co and Mn ions were present in the divalent

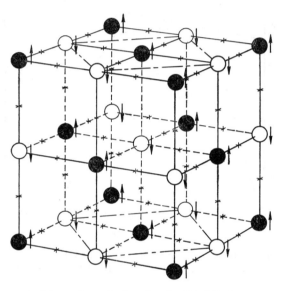

FIG. 7.2. Magnetic ordering in $Ba(B'_{0.5}B''_{0.5})O_3$ type compounds (after Longo and Ward[9]).

and tetravalent valence states respectively in the solid solutions.

Wold *et al.*[7] reported ferromagnetism in the $La(Mn_{1-x}Ni_x)O_3$ system. This may be caused by $Ni^{3+}-Mn^{3+}$ and $Ni^{3+}-Ni^{3+}$ interactions. However, it is possible that the Mn^{4+} and Ni^{2+} may be present in some sites and also could interact.

Sugawara and Iida[8] prepared $BiMnO_3$ at 40 kbar and 700°C. The compound was found to be ferromagnetic at 130°K.

In the magnetic-ordered perovskite-type compounds with the general formula $A(B'_{0.5}B''_{0.5})O_3$ where B' is W^{5+}, Mo^{5+}, or Re^{6+} and B'' also is a parametric ion, Longo and Ward propose that a negative interaction between the B' and B'' exists.[9] The ordered arrangement of the two B cations is shown in Fig. 7.2. The compounds $Ba(Fe_{0.5}Re_{0.5})O_3$, $Sr(Fe_{0.5}Re_{0.5})O_3$ and $Ca(Fe_{0.5}Re_{0.5})O_3$ were found to have Curie temperatures of 43°C, 128°C and 265°C, respectively. By analogy with the

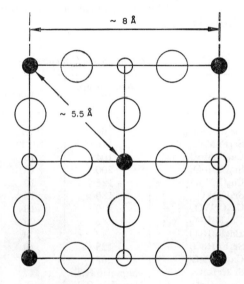

FIG. 7.3. Face of cubic ordered perovskite structure (100). ● = B', ○ = B'', ◯ = 0, A ions omitted (after Blasse[13]).

magnetic properties of these compounds, an increase in Curie temperature with decreasing A cation size might have been anticipated for compounds containing W^{5+} and Mo^{5+} as the B'' ion. However, the $A(Cr_{0.5}Mo_{0.5})O_3$, $A(Cr_{0.5}Mo_{0.5})O_3$ and $A(Fe_{0.5}Mo_{0.5})O_3$ compounds showed a lower Curie temperature for Ca^{2+} compounds.[10] In fact, for the phases in the system $(Ba, Sr, Ca)(Fe_{0.5}Mo_{0.5})O_3$ a maximum in the Curie point was found for $Sr(Fe_{0.5}Mo_{0.5})O_3$ at an Fe–O–Mo bond distance of approximately 3.95 Å.[11]

Blasse, in a study of perovskites $La(B'_{0.5}Mn_{0.5})O_3$ (B' = Mg, Co, Ni and Cu), found that the magnetic exchange interactions between B' and Mn^{4+} ions in ferromagnetic compounds appear to be positive.[12] The saturation moment of $La(Co_{0.5}Mn_{0.5})O_3$ and $La(Ni_{0.5}Mn_{0.5})O_3$ was increased by annealing the compounds, which increased the ordering of the B ions and decreased the number of antiferromagnetic B'–O–B' and Mn^{4+}–O–Mn^{4+} interactions, so that a decreasing Curie temperature was expected and found experimentally. Table 7.1 presents the ferromagnetic perovskite compounds and their Curie points.

TABLE 7.1. *Ferromagnetic Compounds*

	Curie temp. (°C)	References
Ba_2FeMoO_6	64	10, 11
Sr_2FeMoO_6	146	10, 11
Ca_2FeMoO_6	104	10
Sr_2CrMoO_6	200	10
Ca_2CrMoO_6	−125	10
Sr_2CrWO_6	180	10
Ca_2CrWO_6	−130	10
Ba_2FeReO_6	43	9
Ba_2FeReO_6	43	9
Sr_2FeReO_6	128	9
Ca_2FeReO_6	265	9
Sr_2CrReO_6	Magnetic R.T.	17
Ca_2CrReO_6	Magnetic R.T.	17
$BiMnO_3$	−170	8

In a study of antiferromagnetic ordered perovskites $A(B'_{0.5}B''_{0.5})O_3$ where B' is the only paramagnetic ion, Blasse showed that the mechanism for magnetic interaction between the paramagnetic B' ions is of the type B'–O–B''–O–B'[13] (see Fig. 7.3). This is a new type of superexchange in the perovskite structure.

In a study of $Sr(B'_{0.5}W_{0.5})O_3$ type compounds where B' = Mn, Fe, Co and Ni, Blasse[14] found that the Néel temperature (see Fig. 7.4) increased as it did in the $B'O$ compounds

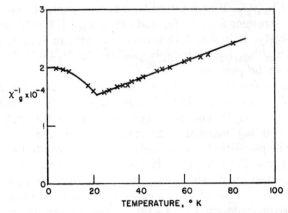

FIG. 7.4. Reciprocal susceptibility per gram vs. temperature of $Sr(Co_{0.5}W_{0.5})O_3$ (after Blasse[14]).

(rocksalt structure) and $KB'F_3$ compounds (perovskite structure). See Table 7.2.

TABLE 7.2. *Magnetic Properties of Compounds* $Sr(B'_{0.5}W_{0.5})O_3$, $B'O$ *and* $KB'F_3$ *(after Blasse[14])*

B'	$Sr(B'_{0.5}W_{0.5})O_3$ T_N (°K)	$B'O$ T_N (°K)	$KB'F_3$ T_N (°K)
Mn^{2+}	10	116	88
Fe^{2+}	16	186	113
Co^{2+}	22	292	114
Ni^{2+}	54	523	275
Cu^{2+}	—	—	243

In the complex perovskite structures the B' ions are separated by an OWO array, and in MeO by a single oxygen ion. Since the Néel temperature is linearly related to the exchange constant of the 180° B'–O–W–O–B' and B'–O–B' interactions, the data in Table 7.2 shows that the B'–O–B' interaction is about 10 times as strong as the B'–O–W–O–B' interaction.

In addition, from measurements on $Ca(Fe_{0.5}Sb_{0.5})O_3$ and

$Sr(Fe_{0.5}Sb_{0.5})O_3$, Blasse found that the magnetic interactions became stronger for shorter distances, T_N (°K) = 31 and 21, respectively. If incomplete ordering exists, such as is found in $Sr(Mn_{0.5}Sb_{0.5})O_3$, a strongly positive $Mn^{3+}-O-Mn^{3+}$ interaction can be present.

Parasitic ferromagnetism has also been observed in perovskites. It was reported in $GdFeO_3$, for temperatures between 78° and 295°K,[15] the magnetization varied for high fields ($H >$ 6000 oe) according to the expression $\sigma = \sigma^0 + \chi H$, where the parasitic ferromagnetization σ^0 amounted to about 1% of the $\sigma(Fe)$ available. It was attributed to imperfectly compensated antiferromagnetism of the Fe^{3+} ion sublattice. Wold[16] has prepared samples of $LaFeO_3$ with reduced parasitic ferromagnetism by careful control of sample stoichiometry, concluding that the use of careful preparation conditions should be a requirement before measurements are made on ferromagnetic materials.

REFERENCES

1. G. H. JONKER and J. H. VAN SANTEN, *Physica* 16, 337 (1950).
2. G. H. JONKER and J. H. VAN SANTEN, *Physica* 19, 120 (1953).
3. G. H. JONKER, *Physica* 22, 707 (1956).
4. J. B. GOODENOUGH, in LANDOLT-BORNSTEIN, *Eigenschaften der Materie in ihren Aggregatszuständen* 9, 2. Springer-Verlag, Berlin (1962).
5. J. B. GOODENOUGH, A. WOLD, R. J. ARNOTT and M. MENYUK, *Phys. Rev.* 124, 373 (1961).
6. G. H. JONKER, *J. Appl. Phys.* 37, 1424 (1966).
7. A. WOLD, R. J. ARNOTT and J. B. GOODENOUGH, *J. Appl. Phys.* 29, 387 (1958).
8. F. SUGAWARA and S. IIDA, *J. Phys. Soc. Japan*, 20, 1529 (1965).
9. J. LONGO and R. WARD, *J. Am. Chem. Soc.* 83, 2816 (1961).
10. F. K. PATTERSON, C. W. MOELLER and R. WARD, *Inorg. Chem.* 2, 196 (1963).
11. F. GALASSO, F. DOUGLAS and R. KASPER, *J. Chem. Phys.* 44, 1672 (1966).
12. G. BLASSE, *J. Phys. Chem. Sol.* 26, 1969 (1965).
13. G. BLASSE, *Proc. Int. Conf. Magnetism*, Nottingham, 350 (1964).
14. G. BLASSE, *Phil. Res. Rpt.* 20, 327 (1965).
15. M. A. GILLEO, *J. Chem. Phys.* 24, 1239 (1956).
16. A. WOLD, private communication.
17. A. W. SLEIGHT, J. LONGO, and R. WARD, *Inorg. Chem.* 1, 245 (1962).

CHAPTER 8

OPTICAL PROPERTIES

8.1. TRANSMITTANCE

Merz[1] studied the optical properties of single-domain crystals of $BaTiO_3$ at various temperatures. The refractive index of the crystal was nearly a constant value of ~ 2.4 from 20° to 90°C and reached a maximum value of 2.46 at 120°C (see Fig. 8.1). At the transition temperature a sudden change in n is observed (see Fig. 8.2).

Lawless and DeVries[2] also measured the index of refraction of $BaTiO_3$ at 5893 Å in the range of 20–160°. A constant index of 2.368 was obtained from 20° to 105°C and above the Curie point the index increased 1.3% to 2.398 and remained constant to 160°C.

The single crystal of $BaTiO_3$, 0.25 mm thick, was found to transmit from 0.5 μ to 6 μ. Complete absorption was found

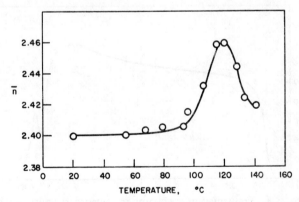

FIG. 8.1. Refractive index $\bar{n}$ as a function of temperature (after Merz[1]).

for wavelengths greater than 11 μ and a feeble absorption band existed near 8 μ (see Fig. 8.3).

The optical properties of strontium titanate single crystals produced by a flame fusion process were reported by Noland.[3] The optical coefficient was obtained from 0.20 μ to 17 μ in wavelength (see Fig. 8.4). Transmission of better than 70% was measured from 0.55 μ to 5 μ. The index of

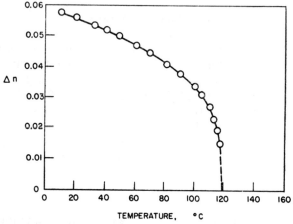

FIG. 8.2. Birefringence Δn as a function of temperature (after Merz[1]).

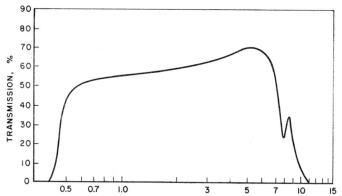

FIG. 8.3. Transmission spectrum of barium titanate single crystal, $d = 0.25$ mm (after A. F. Tatsenko, *Sov. Phys. Techn. Phys.* **3**, 2257 (1957)).

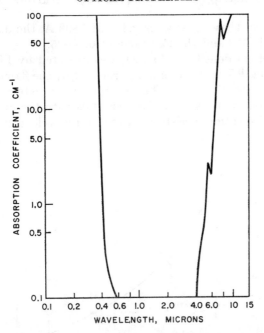

Fig. 8.4. Absorption coefficient of single crystal $SrTiO_3$ as a function of wavelength (after Noland[3]).

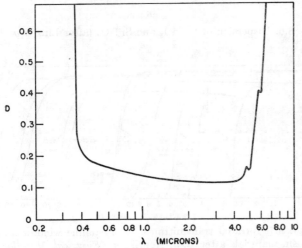

Fig. 8.5. Optical density vs. wavelength of 1.9 mm $CaTiO_3$ crystal (after Linz et al.[4]).

refraction of these crystals is 2.407 at 5893 Å, the dielectric constant is 310 and the loss tangent 0.00025.

The optical density of $CaTiO_3$ was reported by Linz and Herrington.[4] The crystals were prepared by the flame fusion technique (see Fig. 8.5). The absorption characteristics are quite similar to those of $SrTiO_3$ crystals with the exception that the absorptions are shifted to shorter wavelengths. Index of refraction data for $CaTiO_3$ and $SrTiO_3$ as a function of wave-

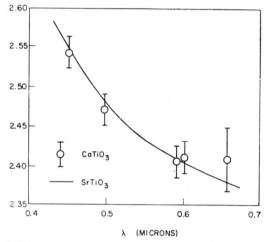

FIG. 8.6. Dispersion of $CaTiO_3$ and $SrTiO_3$ (after Linz et al.[4]).

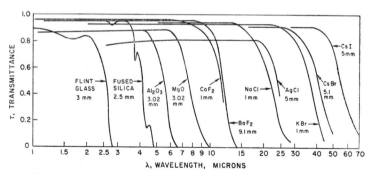

FIG. 8.7. Infrared transmittance of crystalline window and prism materials after L. J. Neuringer *(Electrical Manufacturing)* (March 1960), copyright (Conover-Mast publications, Inc.).

length are plotted in Fig. 8.6. The data are quite similar and the deviation at 657 Å can be attributed to the difficulty in the measuring technique at that wavelength.

$BaTiO_3$ and $SrTiO_3$ have been considered for high-temperature infrared windows. Typical materials which are presently being used are shown in Fig. 8.7. Note that the titanates are useful in the same wavelength range as silica and Al_2O_3. In addition, strontium titanate is considered as an excellent material for use with optically immersed infrared detectors. For many applications the detector–lens combinations are cooled to solid CO_2 and liquid N_2 temperatures to increase the sensitivity. Salzberg[5] made successive transmittance measurements on strontium titanate from room temperature to $-187°C$ showing that there was no decrease in the transmittance of strontium titanate down to $-187°C$.

8.2. Coloration by Light

MacNevin and Ogle[6] noticed that the alkaline earth titanates took on a purplish color when exposed to light and heating reversed the effect. The effect presumably was caused by impurities.

8.3. Electro-optic Effect

The electro-optic properties of $KTaO_3$, $K(Ta_{0.65}Nb_{0.35})O_3$ (KTN), $BaTiO_3$ and $SrTiO_3$ in the paraelectric phase were measured by Geusic et al.[7] The electro-optic coefficients of these perovskites are nearly constant with temperature and from material to material when the distortions of the optical indicatrix are expressed in terms of the induced polarization.[7] Thus, the coefficients might be fundamental properties of the perovskite lattice. These studies also showed that KTN has a large room-temperature electro-optic effect, low electrical losses and a large saturation polarization. An induced birefringence of 0.0057 has been reported with an applied field of 13,000 V/cm, which corresponds to a retardation of 180 half-waves per cm of light path length at 6828 Å. Thus, for polarization of amplitude modulators, the power and voltage requirements can be satisfied by transistor circuitry. The large value of Δn also should make possible a light-scanning system using a prism or a partially electroded cube which can scan over 500 resolvable beam diameters.

Cohen and Gordon[8] described experiments which involve propagating an electromagnetic wave through an electro-optic medium. Using 2.5 μsec microwave pulses with a peak power of a few hundred watts at 9.5 GC, 5% of the light in the zero-order beam was transferred to the higher orders at frequencies 750 GC. Therefore, when the beam is polarized at 45° to the direction of the microwave electric field, it should be possible to extinguish the zero-order beam and pass 75% of the deflected beam.

I. P. Kaminow[9] described experiments in which ferroelectric barium titanate was used at 70 MC as an optical phase modulator. Focused fundamental gaussian mode passed through the edge of a crystal plate with both the higher order "donut" mode of a 0.633-μ He–Ne maser and the fundamental mode maintaining their identity.

Borrelli et al.[10] observed the electro-optic effect of ferroelectric microcrystals in a glass matrix. Crystals of $NaNbO_3$ containing small amounts of cadmium were formed by controlled crystallization of a glass with a high silica glass remaining as a matrix. The cadmium was used to make antiferroelectric $NaNbO_3$ ferroelectric. The crystal size was varied from 50 to 500 Å (less than the wavelength of light) and the dielectric constant from 50 for the glass to 550 for the transparent crystallized material. Materials which had larger crystallite sizes had the higher dielectric constants. In these experiments the difference in refractive index was measured as a function of electric field. No electro-optical activity was observed for the glass, but the retardation was proportional to the square of the electric field for the crystallized glass. In addition, the materials with the higher dielectric constants produce the largest index differences, see Fig. 8.8.

8.4. LASERS

In recent years there has been considerable interest in materials to be used for laser application. The operating laser systems are listed in Table 8.1.

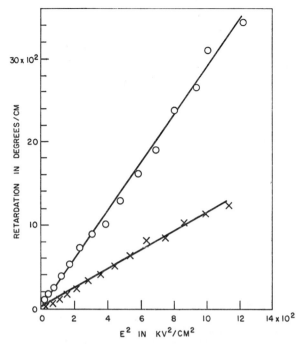

FIG. 8.8. Retardation in degrees of arc per cm vs. the square of the applied electric field (○, sample with ε of 540; ×, sample with ε of 356). Measurements made at 6328 Å (after Borrelli[10]).

In selecting perovskite laser host materials a great deal can be learned from an examination of these systems.

The ion Nd^{3+} appears to be the most popular for introduction into relatively large crystallographic sites. However, except when LaF_3 is used as a host, compensating ions are required in these substitutions. Divalent Tm^{2+} and Dy^{2+} can be substituted in CaF_2 without compensating ions but these divalent rare earth ions are relatively unstable. For Al_2O_3 whose structure contains octahedrally coordinated crystallographic sites for the Al^{3+}, Cr^{3+} proved to be ideal for substitution. Other small ions which have been used to produce laser action are Fe^{3+} and Mn^{4+}, but these ions have not been employed to the extent Cr^{3+} has.

The perovskite structure should be a promising one since

TABLE 8.1. *Data for Operating Laser Systems*

Laser	Output (μ)	Operating temp. (°K)
1. $Al_2O_3(Cr^{3+})$	0.6944	300 (Cont.) 77
2. $CaF_2(U^{3+})$	2.556	77 (Cont.)
3. $SrF_2(U^{3+})$	2.407	77
4. $BaF_2(U^{3+})$	2.556	77
5. Glass (Nd^{3+})	1.06	300
6. $CaWO_4(Nd^{3+})$	1.0646	300 (Cont.)
7. $SrMoO_4(Nd^{3+})$	1.0643 1.064	300 77 (Cont.)
8. $CaMoO_4(Nd^{3+})$	1.067	300
9. $PbMoO_4(Nd^{3+})$	1.0586	300
10. $NaLaMoO_4$ (Nd^{3+})	1.0586	300
11. $CaF_2(Nd^{3+})$	1.046	300
12. $SrF_2(Nd^{3+})$	1.06	300
13. $BaF_2(Nd^{3+})$	1.06	300
14. $LaF_3(Nd^{3+})$	1.06	300
15. $CaWO_4(Ho^{3+})$	2.046	77
16. $CaF_2(Ho^{3+})$	2.05	77
17. $CaWO_4(Tm^{3+})$	1.116	77
18. $SrF_2(Tm^{3+})$	1.91	300
19. $CaWO_4(Fe^{3+})$	1.612	77
20. $CaWO_4(Pr^{3+})$	1.05	77
21. Glass (Yb^{3+})	1.01	77
22. Glass (Sm^{3+})	0.71	77
23. Glass (Gd^{3+})	0.31	
24. $Y_3Al_5O_{12}(Nd^{3+})$		300 (Cont.)
25. $Y_3Al_5O_{12}$ (Nd^{3+}, Cr^{3+})		300 (Cont.)
26. $CaF_2(Sm^{2+})$	0.7082	77
27. $CaF_2(Tm^{2+})$	1.116	4
28. $CaF_2(Dy^{2+})$	2.36	77 (Cont.)

it contains a large A site suitable for Nd^{3+} and smaller B site for Cr^{3+}. In the ideal structure these sites are cubic and centrosymmetric, a condition theoreticians feel should be necessary for obtaining long fluorescence lifetimes for activator ions. However, as was pointed out previously no simple cubic perovskite-type compounds exist which will accept trivalent activator ions without producing defects or require charge compensating ions. The compound $LaAlO_3$, however, has a structure which is only slightly distorted from cubic symmetry. The fluorescence lifetime for Cr^{3+} in $LaAlO_3$ has been reported to be as high as 46 msec as compared with 3 msec for the lifetime of Cr^{3+} in Al_2O_3.[11] This enhancement in lifetime has been attributed to the nearly cubic symmetry of the site containing the Cr^{3+} ion. The transition of $LaAlO_3$ at 435°C presents a problem in obtaining single crystals of thin material. However, flux-grown and hydrothermally grown crystals big enough for fluorescence measurements have been prepared.[12] The measurements showed the fluorescence line at 7356 Å which is characteristic of Cr^{3+}.

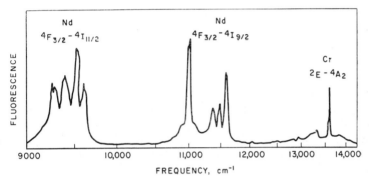

FIG. 8.9. Fluorescence spectrum of $LaAlO_3$–Cr, Nd excited with high-pressure mercury lamp (after Murphy et al.[11]).

Murphy et al.[11] reported on energy transfer between Cr^{3+} and Nd^{3+} in $LaAlO_3$. These studies were conducted on powder samples. The fluorescence spectrum of $LaAlO_3$: Cr, Nd is presented in Fig. 8.9. The line at 7356 Å is due to Cr^{3+} and the remaining two groups are associated with the fluorescence of Nd^{3+}. Lifetimes of Nd^{3+} and Cr^{3+} were found to be 0.46 and 4.6 msec, respectively, at room temperature. Their studies indicated that energy transfer existed between the $^4T_2$ level of Cr^{3+} and the D levels of Nd^{3+}.

Ohlmann and Mazelsky[13] also found transfer of excitation radiation from Cr^{3+} to Nd^{3+} ions in $GdAlO_3$. The purpose of the studies was to increase the efficiency of lasers from the usual 2% or less. The low efficiency was attributed to the poor spectral mismatch between the broad bands of the pump lamps and the relatively narrow absorption bands of the paramagnetic ions. Thus, Cr^{3+} was added to $GdAlO_3$ to produce broad absorption bands which could transfer the energy it absorbed to the Nd^{3+} ion. The fluorescence spectrum showed two groups occurring around 0.9 μ and 1.07 μ from Nd^{3+} and fluorescence at 0.727 μ due to chromium. When the Cr^{3+} levels were excited, the fluorescence of neodymium showed an intensity increase of several times in the doubly doped samples over the singly doped samples. The lifetimes of Nd^{3+} and Cr^{3+} in $GdAlO_3$ were 0.130 msec and 15 msec, respectively.

Other simple perovskite compounds considered for laser host materials were $SrTiO_3$, $SrSnO_3$, $BaSnO_3$ and $BaZrO_3$. However, divalent or tetravalent laser activating ions would be required for substitution so that the symmetry would not be destroyed by compensating ions. The activating ions in these valence states are not particularly stable.

Because of the enhancement of lifetimes observed for Cr^{3+} in $LaAlO_3$ and $GdAlO_3$ as compared with its lifetime in Al_2O_3, a search was made by Galasso et al.[14] for cubic perovskite compounds which would accept trivalent rare earth ions. They found that compounds of the $Ba(B^{3+}_{0.5}Ta_{0.5})O_3$ had an ordered structure in which the B ions alternate so that the symmetry about the B site is retained. Barium was selected since compounds of this type containing barium are usually least distorted, and Ta^{5+} is one of the B ions because of its resistance to reduction and the other B ion should not produce energy levels which would interfere with those of the doping ions. The fluorescence lifetimes of powders of these compounds containing activator ions are presented in Table 8.2. The values obtained, i.e. 850 msec for Nd^{3+} in $Ba(Lu_{0.5}Ta_{0.5})O_3$ and 800 msec in Nd^{3+} in $Ba(Gd_{0.5}Ta_{0.5})O_3$, indicate that there may be some enhancement in lifetimes due to cubic symmetry about the substituted ion. However, the increase is small and it appears that the neodymium lifetime is not influenced as much by environment as that of Cr^{3+}.

The fluorescence spectra of Er^{3+} ion studied in a $Ba(Gd_{0.5}Nb_{0.5})O_3$ host confirmed the fact that a center of symmetry existed at the B site in the structure of these cubic complex oxides.[15]

TABLE 8.2. *Fluorescent Lifetime Data (after Galasso et al.[14])*

Doped $Ba(B_{0.5}Ta_{0.5})O_3$ phases Nd^{3+} doped	Emission line (μ)	Lifetime at 300°K (μ sec)	Lifetime at 77°K (μsec)
$Ba(La_{0.48}Nd_{0.02}Ta_{0.50})O_3$	1.06	500	650
$Ba(Gd_{0.48}Nd_{0.02}Ta_{0.50})O_3$		700	800
$Ba(Y_{0.48}Nd_{0.02}Ta_{0.50})O_3$		200	250
$Ba(Lu_{0.48}Nd_{0.02}Ta_{0.50})O_3$		400	850
$Ba(In_{0.48}Nd_{0.02}Ta_{0.50})O_3$		150	250
$Ba(Sc_{0.48}Nd_{0.02}Ta_{0.50})O_3$		200	220
Sm^{3+} doped			
$Ba(Y_{0.48}Sm_{0.02}Ta_{0.50})O_3$	0.70	1450	
Yb^{3+} doped			
$Ba(Y_{0.48}Yb_{0.02}Ta_{0.50})O_3$	1.01	1800	

REFERENCES

1. W. J. MERZ, *Phys. Rev.* **76**, 1221 (1949).
2. W. N. LAWLESS and R. C. DEVRIES, *J. Appl. Phys.* **35**, 2638 (1964).
3. J. A. NOLAND, *Phys. Rev.* **94**, 724 (1954).
4. A. LINZ, JR. and K. HERRINGTON, *J. Chem. Phys.* **28**, 824 (1958).
5. C. D. SALZBERG, *J. Opt. Soc.* **51**, 1149 (1961).
6. W. M. MACNEVIN and P. R. OGLE, *J. Am. Chem. Soc.* **76**, 3849 (1954).
7. J. E. GEUSIC, S. K. KURTZ, L. G. VAN UITERT and S. H. WEMPLE, *Appl. Phys. Letters* **4**, 14 (1964).
8. M. G. COHEN and E. I. GORDON, *Appl. Phys. Letters* **5**, 181 (1964).
9. I. P. KAMINOW, *Appl. Phys. Letters* **7**, 123 (1965).
10. N. F. BORRELLI, A. HERCZOG and R. D. MAURER, *Appl. Phys. Letters* **7**, 117 (1965).
11. J. MURPHY, R. C. OHLMANN and R. MAZELSKY, *Phys. Rev. Letters* **13**, 135 (1964).
12. Airtron *Semiannual Tech. Sum. Rpt.*, NONR 4616(00) (1 July–31 Dec. 65).
13. R. C. OHLMANN and R. MAZELSKY, to be published.
14. F. GALASSO, G. LAYDEN and D. FLINCHBAUGH, *J. Chem. Phys.* **44**, 2703 (1966).
15. G. BLASSE, A. BRIL and W. C. NIEUWPOORT, *J. Phys. Chem. Soc.* **27**, 1587 (1966).

CHAPTER 9
OTHER PROPERTIES

9.1. CATALYSTS

Fuel Cell

A detailed study was conducted by Epperly et al.[1] to evaluate complex perovskites of the general formula $A(B'_{0.5}B''_{0.5})O_3$ with A = Li, Cs, Ca, Sr, Ba, La, Tl, Pb and Bi, B' = V, Cr, Mn, Fe, Co, Ni, Ru, Rh, Ir and Pt and B'' = Ti, Zr, Sn, Hf, V, Nb, Ta, Mo and W for fuel cell application. The B' ions were selected from those which are known to have catalytic properties and B'' ions were selected for the corrosion resistance they impart to a compound. The problem was to tailor the compounds so that they would have high conductivity and high resistance to attack by dilute solutions of sulfuric acid. While it is dangerous to generalize, the information generated can be quite useful if it is realized that it is qualitative in nature.

They found that most of the perovskite compounds studied were quite acid resistant, but were poor conductors. It is important to note that the B ions were used in their highest oxidation states. The incorporation of oxygen vacancies improved the conductivity but the acid resistance was not as good. Compounds containing Ni had good acid resistance, especially when Ni^{2+} ions were added to the acid, but were poor conductors. The compounds containing cobalt were made conductive by producing oxygen vacancies, but then became soluble in the acid. Compounds containing Nb or Ta were more acid resistant than those containing Mo or W. Compounds containing Ti, Zr or Hf had relatively poor acid resistance. The most satisfactory perovskites were those containing manganese in more than one valence state.

The tungsten bronzes also have been considered for electrodes in fuel cells. Dickens and Whittingham[2] studied the recombination of oxygen atoms on the surfaces of phases with the formula M_xWO_3 where M is Li, Na or K and $0 < x < 0.80$. The pattern of catalytic activity for oxygen atom recombination is approximately the same for Li_xWO_3, Na_xWO_3 and K_xWO_3 phases. In the semiconducting range, from $x = 0$ to $x = 0.25$, the catalytic activity decreased, but in the range above $x = 0.25$ the catalytic activity increased sharply to a maximum at $x = 0.55$. Above $x = 0.25$ the conductivity of the bronzes is metallic. Thus, the catalytic activities of the tungsten bronzes appear to be closely related to the electronic properties of the bronzes.

Heterogeneous Catalysis

Parravano[3] studied the effect of the electronic rearrangement of ferroelectric materials at their transition temperatures on the electron exchange occurring during a catalytic process. From data obtained for carbon monoxide oxidation on $KNbO_3$, he concluded that the change in the rate of catalytic oxidation of CO is affected by the electronic transition. This supported the evidence for an electronic mechanism as being rate determining for the catalytic oxidation of CO.

9.2. THERMAL CONDUCTIVITY

The thermal conductivity values for the titanates are listed in Table 9.1. They are quite low, with the conductivity of barium titanate being lower than that of strontium titanate and calcium titanate at 50°C.

9.3. MELTING POINTS

The melting points of many of the ternary oxides have been determined, but none of complex perovskite type compounds (see Table 9.1). It should be realized that in many cases these are only approximate values; however, they do serve as guides for crystal growth experiments, for example, note that zirconates are extremely high melting-point materials along with hafnates and thorates, the titanates have

TABLE 9.1. *Other Properties*

Phase	Thermal conductivity (W/cm°C)	Coeff. of thermal Expansion ($\times 10^{-6}/°C$)		ΔH (kcal/mole)	Melting point (°C)
$BaThO_3$					2299
$BaTiO_3$	0.022–0.032 (50–230)	19 14	(10–90) (140–1200)	394.8	1610
$BaZrO_3$		7.79 8.52	(25–500) (25–1000)		2688
$CaHfO_3$		3.6 12.1 7.08	(100–600) (800–1300) (1000–1300)		2471
$CaTiO_3$	0.043–0.046 (50–135)	13.04 14.05	(25–500) (25–1000)	397.4	1975
$CaZrO_3$					2343
$GdAlO_3$					2030
$KNbO_3$					1039
$KTaO_3$					1370
$LaAlO_3$					2075–2080
$LaFeO_3$					1888
$NaNbO_3$					1412
$NaTaO_3$					1780
$PbTiO_3$		−16 (below 490) 25 (above 490°)			
$SrTiO_3$	0.037–0.058 (50–140)	8.63 9.43		398.9	2080
$SrZrO_3$		8.84 9.64	(25–500) (25–1000)		2799
$YAlO_3$					1950
$Ba(Fe_{0.5}Ta_{0.5})O_3$		13.19			
$Sr(Fe_{0.5}Ta_{0.5})O_3$		10.99			

intermediate melting points, and the niobates and tantalates are relatively low melting.

9.4. HEATS OF FORMATION

Only the heats of formation of the titanates have been measured. These values shown in Table 9.1 are quite similar for barium, strontium and calcium titanate.

9.5. Thermal Expansion

There does not appear to be a trend in the thermal expansion of the perovskites in relation to the ion in the A position (see Table 9.1). Note that for the titanates, the coefficient of thermal expansion of barium titanate above the Curie point is similar to that of calcium titanate and larger than that of strontium titanate. For the zirconates the coefficient of thermal expansion of $SrZrO_3$ is greater than that for $BaZrO_3$, while the coefficient of thermal expansion of $Ba(Fe_{0.5}Ta_{0.5})O_3$ is greater than that of $Sr(Fe_{0.5}Ta_{0.5})O_3$. Lead titanate is unusual, having a negative coefficient up to 490°C and above this temperature it behaves like the other titanates.

9.6. Density

The densities of perovskite-type compounds were calculated from X-ray data using the equation in a computer program

$$D = \frac{\text{Molecular weight} \times \text{no. of molecules per unit cell}}{\text{Volume of unit cell} \times 6.023 \times 10^{23} \text{ molecules/mole}}$$

In the program, the equation for the volume of a triclinic cell was used

$$V = a \times b \times c \sqrt{(1 - \cos^2 \alpha - \cos^2 \beta - \cos^2 \gamma + 2 \cos \alpha \cos \beta \cos \gamma)}$$

and the appropriate parameters from Table 2.2 were introduced for each compound. The molecular weight was obtained by introducing the periodic table with atomic weights into the computer and then inserting the data deck with the appropriate atomic symbols and multipliers. The cube root of

$$\frac{\text{volume of the unit cell}}{\text{no. of molecules per unit cell}}$$

was also calculated to obtain a', the unit cell edge of a cubic cell which represents the same volume allowed in the true unit cell for a molecule of a perovskite-type compound. This permits a comparison to be made between the ionic radii of the ions in these compounds even when they are indexed on cells with different symmetry. Table 9.2 presents the molecular weight, a', and the density of the perovskite compounds.

9.7. Mechanical Properties

Only very little information is available on the mechanical properties of perovskites (see Table 9.1). The modulus of barium titanate has been measured to be 16×10^6 psi.

The modulus of $SrZrO_3$ has been found to be 12×10^6 psi, the bend strength 13×10^3 psi and the compressive strength 5×10^3 psi. Strontium zirconate has an interesting stress–strain curve as shown in Fig. 9.1.[4] The small yield observed was attributed to domain motion or crystal twinning similar to that which takes place in ferroelectric $BaTiO_3$ under an electric stress.

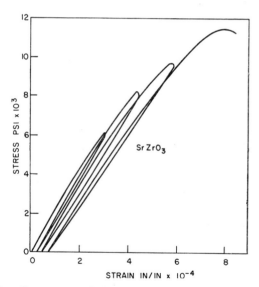

Fig. 9.1. Stress–strain behavior of hot-pressed $SrZrO_3$ (after Tinklepaugh[4]).

TABLE 9.2. *Data for Perovskite-type Compounds*

Compound	Formula weight	Volume	Z	a', Å	Density
$A^{1+}B^{5+}O_3$					
$AgNbO_3$	248.774	974.325	16	3.934	6.782
$AgTaO_3$	336.816	60.481	1	3.925	9.244
$CsIO_3$	307.808	101.195	1	4.660	5.049
KIO_3	214.005	85.752	1	4.410	4.143
$KNbO_3$	180.006	129.368	2	4.014	4.620
$KTaO_3$	268.048	63.450	1	3.988	7.013
$NaNbO_3$	163.894	119.427	2	3.909	4.556
$NaTaO_3$	251.936	117.498	2	3.887	7.119
$RbIO_3$	260.373	92.345	1	4.520	4.680
$TlIO_3$	379.273	91.716	1	4.510	6.865
$A^{2+}B^{4+}O_3$					
$BaCeO_3$	325.458	85.010	1	4.397	6.355
$BaFeO_3$	241.185	63.520	1	3.990	6.303
$BaMoO_3$	281.278	65.959	1	4.040	7.079
$BaPbO_3$	392.528	78.019	1	4.273	8.352
$BaPrO_3$	326.245	82.540	1	4.354	6.561
$BaPuO_3$	427.338	84.605	1	4.390	8.385
$BaSnO_3$	304.028	69.782	1	4.117	7.232
$BaThO_3$	417.376	90.217	1	4.485	7.680
$BaTiO_3$	233.238	64.110	1	4.002	6.039
$BaUO_3$	423.368	84.431	1	4.387	8.324
$BaZrO_3$	276.558	73.665	1	4.192	6.232
$CaCeO_3$	228.198	460.683	8	3.862	6.578
$CaHfO_3$	266.568	—	—	—	—
$CaMnO_3$	143.016	833.656	8	4.706	2.278
$CaMoO_3$	184.018	472.590	8	3.895	5.171
$CaSnO_3$	206.768	246.406	4	3.949	5.572
$CaThO_3$	320.116	667.628	8	4.370	6.368
$CaTiO_3$	135.978	223.913	4	3.825	4.032
$CaUO_3$	326.108	286.060	4	4.151	7.570
$CaVO_3$	139.020	215.125	4	3.775	4.291
$CaZrO_3$	179.298	257.617	4	4.008	4.621
$CdCeO_3$	300.518	447.697	8	3.825	8.914
$CdSnO_3$	279.088	243.371	4	3.933	7.615
$CdThO_3$	392.436	667.628	8	4.370	7.806
$CdTiO_3$	208.298	218.491	4	3.794	6.330
$CdZrO_3$	251.618	—	—	—	—
$EuTiO_3$	247.858	59.182	1	3.897	6.952
$MgCeO_3$	212.430	622.836	8	4.270	4.529

TABLE 9.2 *(cont.)*

Compound	Formula weight	Volume	Z	a', Å	Density
$A^{2+}B^{4+}O_3$ *(cont.)*					
$PbCeO_3$	395.308	442.451	8	3.810	11.865
$PbHfO_3$	433.678	—	—	—	—
$PbSnO_3$	373.878	502.268	8	3.974	9.885
$PbTiO_3$	303.088	62.780	1	3.974	8.014
$PbZrO_3$	346.408	799.179	8	4.640	5.756
$SrCeO_3$	275.738	312.783	4	4.276	5.854
$SrCoO_3$	194.551	460.994	8	3.862	5.605
$SrFeO_3$	191.465	57.916	1	3.869	5.488
$SrHfO_3$	314.108	67.369	1	4.069	7.740
$SrMoO_3$	231.558	62.812	1	3.975	6.120
$SrPbO_3$	342.808	290.801	4	4.174	7.828
$SrRuO_3$	236.688	—	—	—	—
$SrSnO_3$	254.308	65.617	1	4.033	6.434
$SrThO_3$	367.656	690.807	8	4.420	7.068
$SrTiO_3$	183.518	59.502	1	3.904	5.120
$SrUO_3$	373.648	318.903	4	4.304	7.780
$SrZrO_3$	226.838	275.952	4	4.101	5.458
$A^{3+}B^{3+}O_3$					
$BiAlO_3$	283.960	459.822	8	3.859	8.201
$BiCrO_3$	308.974	58.851	1	3.890	8.715
$BiMnO_3$	311.916	61.424	1	3.946	8.430
$CeAlO_3$	215.100	53.838	1	3.776	6.632
$CeCrO_3$	240.114	57.781	1	3.866	6.898
$CeFeO_3$	243.965	59.319	1	3.900	6.827
$CeGaO_3$	257.838	58.366	1	3.879	7.333
$CeScO_3$	233.074	—	—	—	—
$CeVO_3$	239.060	59.319	1	3.900	6.690
$CrBiO_3$	308.974	487.813	8	3.936	8.411
$DyAlO_3$	237.480	204.722	4	3.713	7.702
$DyFeO_3$	266.345	226.162	4	3.838	7.820
$DyMnO_3$	265.436	50.653	1	3.700	8.699
$EuAlO_3$	226.940	208.034	4	3.733	7.243
$EuCrO_3$	251.954	55.002	1	3.803	7.604
$EuFeO_3$	255.805	231.631	4	3.869	7.333
$FeBiO_3$	312.825	62.335	1	3.965	8.331
$GdAlO_3$	232.230	207.251	4	3.728	7.440
$GdCoO_3$	264.181	52.228	1	3.738	8.397
$GdCrO_3$	257.244	222.929	4	3.820	7.662
$GdFeO_3$	261.095	230.217	4	3.861	7.531
$GdMnO_3$	260.186	55.743	1	3.820	7.748

TABLE 9.2 (cont.)

Compound	Formula weight	Volume	Z	a', Å	Density
$A^{3+}B^{3+}O_3$ (cont.)					
$GdScO_3$	250.204	250.297	4	3.970	6.638
$GdVO_3$	256.190	229.560	4	3.857	7.410
$LaAlO_3$	213.890	54.354	1	3.788	6.532
$LaCoO_3$	245.841	55.906	1	3.824	7.300
$LaCrO_3$	238.904	234.202	4	3.883	6.773
$LaFeO_3$	242.755	243.086	4	3.932	6.631
$LaGaO_3$	256.628	236.413	4	3.895	7.208
$LaInO_3$	186.908	277.773	4	4.110	4.468
$LaNiO_3$	245.618	452.246	8	3.838	7.212
$LaRhO_3$	289.813	61.163	1	3.940	7.866
$LaScO_3$	231.864	266.089	4	4.052	5.786
$LaTiO_3$	234.808	60.236	1	3.920	6.471
$LaVO_3$	237.850	63.521	1	3.990	6.210
$LaYO_3$	275.813	—	—	—	—
$NdAlO_3$	219.220	52.819	1	3.752	6.890
$NdCoO_3$	251.171	53.882	1	3.777	7.738
$NdCrO_3$	244.234	228.799	4	3.853	7.088
$NdFeO_3$	248.085	235.092	4	3.888	7.007
$NdGaO_3$	261.958	230.054	4	3.860	7.561
$NdInO_3$	192.238	269.200	4	4.068	4.742
$NdMnO_3$	247.176	54.872	1	3.800	7.478
$NdScO_3$	237.194	257.276	4	4.007	6.122
$NdVO_3$	243.180	235.115	4	3.888	6.868
$PrAlO_3$	215.887	37.778	1	3.355	9.486
$PrCoO_3$	247.838	54.310	1	3.787	7.575
$PrCrO_3$	240.901	230.181	4	3.861	6.949
$PrFeO_3$	244.752	239.385	4	3.912	6.789
$PrGaO_3$	258.625	232.103	4	3.871	7.399
$PrMnO_3$	243.843	55.743	1	3.820	7.262
$PrScO_3$	233.861	260.334	4	4.022	5.965
$PrVO_3$	239.847	235.641	4	3.891	6.759
$PuAlO_3$	316.980	97.324	1	4.600	5.407
$PuCrO_3$	341.994	233.456	4	3.879	9.727
$PuMnO_3$	344.936	57.512	1	3.860	9.956
$PuVO_3$	340.940	239.179	4	3.910	9.465
$SmAlO_3$	225.330	208.928	4	3.738	7.161
$SmCoO_3$	257.281	53.123	1	3.759	8.040
$SmCrO_3$	250.344	226.109	4	3.838	7.352
$SmFeO_3$	254.195	232.589	4	3.874	7.257
$SmInO_3$	198.348	265.872	4	4.051	4.954
$SmVO_3$	249.290	58.864	1	3.890	7.030
$YAlO_3$	163.885	203.404	4	3.705	5.350

TABLE 9.2 *(cont.)*

Compound	Formula weight	Volume	Z	a', Å	Density
$A^{3+}B^{3+}O_3$ *(cont.)*					
$YCrO_3$	188.899	218.305	4	3.793	5.746
$YFeO_3$	192.750	225.862	4	3.836	5.667
$YScO_3$	181.859	244.887	4	3.941	4.931
A_xBO_3					
$Ce_{0.33}NbO_3$	187.144	119.550	2	3.910	5.197
$Ce_{0.33}TaO_3$	275.186	119.857	2	3.913	7.623
$Dy_{0.33}TaO_3$	282.571	113.684	2	3.845	8.252
$Gd_{0.33}TaO_3$	280.839	116.370	2	3.875	8.012
$La_{0.33}NbO_3$	186.744	120.776	2	3.923	5.133
$La_{0.33}TaO_3$	274.786	121.087	2	3.927	7.534
$Nd_{0.33}NbO_3$	188.503	118.332	2	3.897	5.289
$Nd_{0.33}TaO_3$	276.545	118.789	2	3.902	7.729
$Pr_{0.33}NbO_3$	187.404	119.092	2	3.905	5.224
$Pr_{0.33}TaO_3$	275.446	119.246	2	3.907	7.669
$Sm_{0.33}TaO_3$	278.562	117.274	2	3.885	7.886
$Y_{0.33}TaO_3$	258.285	113.227	2	3.840	7.573
$Yb_{0.33}TaO_3$	286.049	110.852	2	3.813	8.567
$Ca_{0.5}TaO_3$	248.986	446.725	4	4.816	3.701
Li_xWO_3 $x=1$	238.787	51.479	1	3.720	7.700
Na_xWO_3	254.838	57.602	1	3.862	7.344
$Sr_{0.7}NbO_3$	202.238	63.092	1	3.981	5.321
$Sr_{0.95}NbO_3$	224.143	64.771	1	4.016	5.745
$CaMnO_{3-x}$ $x=6$	143.000	—	—	—	—
$SrCoO_{3-x}$ $x=0$	194.535	—	—	—	—
$SrFeO_{3-x}$ $x=0$	191.449	—	—	—	—
$SrTiO_{3-x}$ $x=0$	183.502	—	—	—	—
$SrVO_{3-x}$ $x=0$	186.544	—	—	—	—
$A(B'_{0.67}B''_{0.33})O_3$					
$Ba(Al_{0.67}W_{0.33})O_3$	264.086	—	—	—	—
$Ba(Dy_{0.67}W_{0.33})O_3$	354.884	589.745	8	4.193	7.991
$Ba(Er_{0.67}W_{0.33})O_3$	358.073	589.745	8	4.193	8.063
$Ba(Eu_{0.67}W_{0.33})O_3$	347.822	637.166	8	4.302	7.249
$Ba(Fe_{0.67}U_{0.33})O_3$	301.306	557.848	8	4.116	7.173
$Ba(Gd_{0.67}W_{0.33})O_3$	351.366	595.036	8	4.205	7.842
$Ba(In_{0.67}W_{0.33})O_3$	246.009	576.138	8	4.160	5.671
$Ba(La_{0.67}W_{0.33})O_3$	339.078	631.629	8	4.290	7.129
$Ba(In_{0.67}U_{0.33})O_3$	263.888	616.730	8	4.256	5.682
$Ba(Lu_{0.67}W_{0.33})O_3$	363.239	—	—	—	—
$Ba(Nd_{0.67}W_{0.33})O_3$	342.649	616.947	8	4.256	7.376
$Ba(Sc_{0.67}U_{0.33})O_3$	294.009	611.960	8	4.245	6.380

TABLE 9.2 (cont.)

Compound	Formula weight	Volume	Z	a', Å	Density
$A(B'_{0.67}B''_{0.33})O_3$ (cont.)					
$Ba(Sc_{0.67}W_{0.33})O_3$	276.129	559.476	8	4.120	6.554
$Ba(Y_{0.67}U_{0.33})O_3$	323.454	658.503	8	4.350	6.523
$Ba(Y_{0.67}W_{0.33})O_3$	305.575	587.217	8	4.187	6.911
$Ba(Yb_{0.67}W_{0.33})O_3$	361.945	—	—	—	—
$Pb(Fe_{0.67}W_{0.33})O_3$	353.276	—	—	—	—
$Sr(Cr_{0.67}Re_{0.33})O_3$	231.902	513.922	8	4.005	5.992
$Sr(Cr_{0.67}U_{0.33})O_3$	249.005	512.000	8	4.000	6.459
$Sr(Fe_{0.67}Re_{0.33})O_3$	234.482	491.169	8	3.945	6.340
$Sr(Fe_{0.67}W_{0.33})O_3$	233.706	61.490	1	3.947	6.309
$Sr(In_{0.67}Re_{0.33})O_3$	197.064	571.167	8	4.148	4.582
$La(Co_{0.67}Nb_{0.33})O_3$	257.052	245.666	4	3.945	6.948
$La(Co_{0.67}Sb_{0.33})O_3$	266.571	244.166	4	3.937	7.249
$A(B'_{0.33}B''_{0.67})O_3$					
$Ba(Ca_{0.33}Nb_{0.67})O_3$	260.812	220.045	3	4.186	5.903
$Ba(Ca_{0.33}Ta_{0.67})O_3$	319.800	219.214	3	4.181	7.265
$Ba(Cd_{0.33}Nb_{0.67})O_3$	284.677	72.407	1	4.168	6.526
$Ba(Cd_{0.33}Ta_{0.67})O_3$	343.665	72.355	1	4.167	7.884
$Ba(Co_{0.33}Nb_{0.67})O_3$	267.033	68.418	1	4.090	6.479
$Ba(Co_{0.33}Ta_{0.67})O_3$	326.021	204.617	3	4.086	7.935
$Ba(Cu_{0.33}Nb_{0.67})O_3$	268.553	542.989	8	4.079	6.568
$Ba(Fe_{0.33}Nb_{0.67})O_3$	266.015	68.167	1	4.085	6.478
$Ba(Fe_{0.33}Ta_{0.67})O_3$	325.003	68.921	1	4.100	7.828
$Ba(Mg_{0.33}Nb_{0.67})O_3$	255.608	204.134	3	4.083	6.236
$Ba(Mg_{0.33}Ta_{0.67})O_3$	314.596	204.608	3	4.086	7.657
$Ba(Mn_{0.33}Nb_{0.67})O_3$	265.715	—	—	—	—
$Ba(Mn_{0.33}Ta_{0.67})O_3$	324.703	208.994	3	4.115	7.737
$Ba(Ni_{0.33}Nb_{0.67})O_3$	266.960	67.618	1	4.074	6.554
$Ba(Ni_{0.33}Ta_{0.67})O_3$	325.948	202.482	3	4.072	8.017
$Ba(Pb_{0.33}Nb_{0.67})O_3$	315.958	77.309	1	4.260	6.784
$Ba(Pb_{0.33}Ta_{0.67})O_3$	374.946	76.766	1	4.250	8.108
$Ba(Sr_{0.33}Ta_{0.67})O_3$	335.488	229.026	3	4.242	7.295
$Ba(Zn_{0.33}Nb_{0.67})O_3$	269.157	68.619	1	4.094	6.511
$Ba(Zn_{0.33}Ta_{0.67})O_3$	328.145	205.476	3	4.091	7.953
$Ca(Ni_{0.33}Nb_{0.67})O_3$	169.700	58.411	1	3.880	4.823
$Ca(Ni_{0.33}Ta_{0.67})O_3$	228.688	60.698	1	3.930	6.254
$Pb(Co_{0.33}Nb_{0.67})O_3$	336.883	65.939	1	4.040	8.481
$Pb(Co_{0.33}Ta_{0.67})O_3$	395.871	64.481	1	4.010	10.191
$Pb(Mg_{0.33}Nb_{0.67})O_3$	325.458	65.988	1	4.041	8.187
$Pb(Mg_{0.33}Ta_{0.67})O_3$	384.446	64.965	1	4.020	8.823
$Pb(Mn_{0.33}Nb_{0.67})O_3$	335.565	—	—	—	—
$Pb(Ni_{0.33}Nb_{0.67})O_3$	336.810	65.208	1	4.025	8.574

150 PREPARATION OF PEROVSKITE-TYPE COMPOUNDS

TABLE 9.2 *(cont.)*

Compound	Formula weight	Volume	Z	a', Å	Density
$A(B'_{0.33}B''_{0.67})O_3$ *(cont.)*					
$Pb(Ni_{0.33}Ta_{0.67})O_3$	395.798	64.481	1	4.010	10.189
$Pb(Zn_{0.33}Nb_{0.67})O_3$	339.007	65.939	1	4.040	8.534
$Sr(Ca_{0.33}Nb_{0.67})O_3$	211.092	205.726	3	4.093	5.110
$Sr(Ca_{0.33}Sb_{0.67})O_3$	230.417	545.338	8	4.085	5.611
$Sr(Ca_{0.33}Ta_{0.67})O_3$	270.080	204.170	3	4.083	6.588
$Sr(Cd_{0.33}Nb_{0.67})O_3$	234.957	68.368	1	4.089	5.705
$Sr(Co_{0.33}Nb_{0.67})O_3$	217.313	513.922	8	4.005	5.615
$Sr(Co_{0.33}Sb_{0.67})O_3$	236.639	510.082	8	3.995	6.161
$Sr(Co_{0.33}Ta_{0.67})O_3$	276.301	190.423	3	3.989	7.226
$Sr(Cu_{0.33}Sb_{0.67})O_3$	238.159	503.403	8	3.977	6.283
$Sr(Fe_{0.33}Nb_{0.67})O_3$	216.295	64.192	1	4.004	5.593
$Sr(Mg_{0.33}Nb_{0.67})O_3$	205.888	193.651	3	4.011	5.295
$Sr(Mg_{0.33}Sb_{0.67})O_3$	225.214	504.358	8	3.980	5.930
$Sr(Mg_{0.33}Ta_{0.67})O_3$	264.876	192.301	3	4.002	6.859
$Sr(Mn_{0.33}Nb_{0.67})O_3$	215.995	—	—	—	—
$Sr(Mn_{0.33}Ta_{0.67})O_3$	274.983	—	—	—	—
$Sr(Ni_{0.33}Nb_{0.67})O_3$	217.240	190.081	3	3.987	5.692
$Sr(Ni_{0.33}Ta_{0.67})O_3$	276.228	188.489	3	3.975	7.298
$Sr(Pb_{0.33}Nb_{0.67})O_3$	266.238	—	—	—	—
$Sr(Pb_{0.33}Ta_{0.67})O_3$	325.226	—	—	—	—
$Sr(Zn_{0.33}Nb_{0.67})O_3$	219.437	192.818	3	4.006	5.667
$Sr(Zn_{0.33}Ta_{0.67})O_3$	278.425	193.119	3	4.008	7.180
$A(B^{3+}_{0.5}B^{5+}_{0.5})O_3$					
$Ba(Bi_{0.5}Nb_{0.5})O_3$	336.281	642.736	8	4.315	6.948
$Ba(Bi_{0.5}Ta_{0.5})O_3$	380.302	628.982	8	4.284	8.030
$Ba(Ce_{0.5}Nb_{0.5})O_3$	301.851	79.119	1	4.293	6.333
$Ba(Ce_{0.5}Pa_{0.5})O_3$	370.898	681.472	8	4.400	7.228
$Ba(Co_{0.5}Nb_{0.5})O_3$	261.258	66.923	1	4.060	6.480
$Ba(Co_{0.5}Re_{0.5})O_3$	307.905	528.690	8	4.043	7.734
$Ba(Cr_{0.5}Os_{0.5})O_3$	306.436	—	—	—	—
$Ba(Cr_{0.5}Re_{0.5})O_3$	304.436	—	—	—	—
$Ba(Cr_{0.5}U_{0.5})O_3$	330.351	—	—	—	—
$Ba(Cu_{0.5}W_{0.5})O_3$	309.033	534.633	8	4.058	7.676
$Ba(Dy_{0.5}Nb_{0.5})O_3$	313.041	600.571	8	4.218	6.922
$Ba(Dy_{0.5}Pa_{0.5})O_3$	382.088	667.628	8	4.370	7.600
$Ba(Dy_{0.5}Ta_{0.5})O_3$	357.062	623.930	8	4.272	7.600
$Ba(Er_{0.5}Nb_{0.5})O_3$	315.421	598.438	8	4.213	7.000
$Ba(Er_{0.5}Pa_{0.5})O_3$	384.468	662.143	8	4.358	7.711
$Ba(Er_{0.5}Re_{0.5})O_3$	362.068	583.020	8	4.177	8.247
$Ba(Er_{0.5}Ta_{0.5})O_3$	359.442	597.586	8	4.211	7.988
$Ba(Er_{0.5}U_{0.5})O_3$	387.983	651.714	8	4.335	7.906

TABLE 9.2 *(cont.)*

Compound	Formula weight	Volume	Z	a', Å	Density
$A(B^{3+}_{0.5}B^{5+}_{0.5})O_3$ *(cont.)*					
$Ba(Eu_{0.5}Nb_{0.5})O_3$	307.771	615.643	8	4.253	6.639
$Ba(Eu_{0.5}Pa_{0.5})O_3$	376.818	677.530	8	4.391	7.386
$Ba(Eu_{0.5}Ta_{0.5})O_3$	351.792	615.426	8	4.253	7.591
$Ba(Fe_{0.5}Mo_{0.5})O_3$	261.232	527.514	8	4.040	6.576
$Ba(Fe_{0.5}Nb_{0.5})O_3$	259.715	66.923	1	4.060	6.442
$Ba(Fe_{0.5}Re_{0.5})O_3$	306.362	521.660	8	4.025	7.799
$Ba(Fe_{0.5}Ta_{0.5})O_3$	303.736	66.726	1	4.056	7.556
$Ba(Gd_{0.5}Nb_{0.5})O_3$	310.416	613.258	8	4.248	6.722
$Ba(Gd_{0.5}Pa_{0.5})O_3$	379.463	675.449	8	4.387	7.461
$Ba(Gd_{0.5}Re_{0.5})O_3$	357.063	599.290	8	4.215	7.912
$Ba(Gd_{0.5}Sb_{0.5})O_3$	324.838	601.212	8	4.220	7.175
$Ba(Gd_{0.5}Ta_{0.5})O_3$	354.437	613.184	8	4.248	7.676
$Ba(Ho_{0.5}Nb_{0.5})O_3$	314.256	599.930	8	4.217	6.956
$Ba(Ho_{0.5}Pa_{0.5})O_3$	383.303	665.339	8	4.365	7.651
$Ba(Ho_{0.5}Ta_{0.5})O_3$	358.277	601.639	8	4.221	7.908
$Ba(In_{0.5}Nb_{0.5})O_3$	231.791	567.458	8	4.139	5.425
$Ba(In_{0.5}Os_{0.5})O_3$	280.438	556.223	8	4.112	6.696
$Ba(In_{0.5}Pa_{0.5})O_3$	300.838	635.169	8	4.298	6.290
$Ba(In_{0.5}Re_{0.5})O_3$	278.438	563.151	8	4.129	6.566
$Ba(In_{0.5}Sb_{0.5})O_3$	246.213	565.404	8	4.134	5.783
$Ba(In_{0.5}Ta_{0.5})O_3$	275.812	567.664	8	4.140	6.452
$Ba(In_{0.5}U_{0.5})O_3$	304.353	618.470	8	4.260	6.535
$Ba(La_{0.5}Nb_{0.5})O_3$	301.246	643.759	8	4.317	6.214
$Ba(La_{0.5}Pa_{0.5})O_3$	370.293	701.411	8	4.442	7.011
$Ba(La_{0.5}Re_{0.5})O_3$	347.893	631.629	8	4.290	7.314
$Ba(La_{0.5}Ta_{0.5})O_3$	345.267	651.958	8	4.336	7.033
$Ba(Lu_{0.5}Nb_{0.5})O_3$	319.276	585.116	8	4.182	7.246
$Ba(Lu_{0.5}Pa_{0.5})O_3$	388.323	650.813	8	4.333	7.924
$Ba(Lu_{0.5}Ta_{0.5})O_3$	363.297	586.797	8	4.186	8.222
$Ba(Mn_{0.5}Nb_{0.5})O_3$	259.260	68.067	1	4.083	6.323
$Ba(Mn_{0.5}Re_{0.5})O_3$	305.907	547.343	8	4.090	7.422
$Ba(Mn_{0.5}Ta_{0.5})O_3$	255.283	67.718	1	4.076	6.258
$Ba(Nd_{0.5}Nb_{0.5})O_3$	303.911	622.836	8	4.270	6.480
$Ba(Nd_{0.5}Pa_{0.5})O_3$	372.958	690.807	8	4.420	7.170
$Ba(Nd_{0.5}Re_{0.5})O_3$	350.558	616.295	8	4.255	7.554
$Ba(Nd_{0.5}Ta_{0.5})O_3$	347.932	626.343	8	4.278	7.377
$Ba(Ni_{0.5}Nb_{0.5})O_3$	261.146	68.921	1	4.100	6.290
$Ba(Pr_{0.5}Nb_{0.5})O_3$	302.245	77.854	1	4.270	6.444
$Ba(Pr_{0.5}Pa_{0.5})O_3$	371.292	695.978	8	4.431	7.085
$Ba(Pr_{0.5}Ta_{0.5})O_3$	346.266	77.854	1	4.270	7.383
$Ba(Rh_{0.5}Nb_{0.5})O_3$	283.244	545.338	8	4.085	6.898
$Ba(Rh_{0.5}U_{0.5})O_3$	355.806	—	—	—	—

TABLE 9.2 (cont.)

Compound	Formula weight	Volume	Z	a', Å	Density
$A(B^{3+}_{0.5}B^{5+}_{0.5})O_3$ (cont.)					
$Ba(Sc_{0.5}Nb_{0.5})O_3$	254.269	69.985	1	4.121	6.031
$Ba(Sc_{0.5}Os_{0.5})O_3$	302.916	541.742	8	4.076	7.426
$Ba(Sc_{0.5}Pa_{0.5})O_3$	323.316	624.807	8	4.274	6.872
$Ba(Sc_{0.5}Re_{0.5})O_3$	300.916	543.938	8	4.081	7.347
$Ba(Sc_{0.5}Sb_{0.5})O_3$	268.691	550.763	8	4.098	6.479
$Ba(Sc_{0.5}Ta_{0.5})O_3$	298.290	555.818	8	4.111	7.127
$Ba(Sc_{0.5}U_{0.5})O_3$	326.831	611.960	8	4.245	7.092
$Ba(Sm_{0.5}Nb_{0.5})O_3$	306.966	618.035	8	4.259	6.596
$Ba(Sm_{0.5}Pa_{0.5})O_3$	376.013	679.615	8	4.396	7.347
$Ba(Sm_{0.5}Ta_{0.5})O_3$	350.987	618.252	8	4.259	7.539
$Ba(Tb_{0.5}Nb_{0.5})O_3$	311.253	75.633	1	4.229	6.831
$Ba(Tb_{0.5}Pa_{0.5})O_3$	380.300	670.611	8	4.376	7.531
$Ba(Tl_{0.5}Ta_{0.5})O_3$	377.997	596.948	8	4.210	8.409
$Ba(Tm_{0.5}Nb_{0.5})O_3$	316.258	594.399	8	4.204	7.066
$Ba(Tm_{0.5}Pa_{0.5})O_3$	385.305	656.688	8	4.346	7.792
$Ba(Tm_{0.5}Ta_{0.5})O_3$	360.279	593.975	8	4.203	8.055
$Ba(Y_{0.5}Nb_{0.5})O_3$	276.244	74.088	1	4.200	6.189
$Ba(Y_{0.5}Pa_{0.5})O_3$	345.291	662.599	8	4.359	6.920
$Ba(Y_{0.5}Re_{0.5})O_3$	322.891	586.797	8	4.186	7.307
$Ba(Y_{0.5}Ta_{0.5})O_5$	320.265	599.717	8	4.216	7.092
$Ba(Y_{0.5}U_{0.5})O_3$	348.806	656.235	8	4.345	7.059
$Ba(Yb_{0.5}Nb_{0.5})O_3$	318.311	587.217	8	4.187	7.199
$Ba(Yb_{0.5}Pa_{0.5})O_3$	387.358	653.520	8	4.339	7.871
$Ba(Yb_{0.5}Ta_{0.5})O_3$	362.332	590.590	8	4.195	8.147
$Ca(Al_{0.5}Nb_{0.5})O_3$	148.022	55.161	1	3.807	4.455
$Ca(Al_{0.5}Ta_{0.5})O_3$	192.043	55.161	1	3.807	5.779
$Ca(Co_{0.5}W_{0.5})O_3$	209.470	235.054	4	3.888	5.917
$Ca(Cr_{0.5}Mo_{0.5})O_3$	162.046	226.583	4	3.841	4.749
$Ca(Cr_{0.5}Nb_{0.5})O_3$	160.529	57.061	1	3.850	4.670
$Ca(Cr_{0.5}Os_{0.5})O_3$	209.176	225.423	4	3.834	6.161
$Ca(Cr_{0.5}Re_{0.5})O_3$	207.176	225.717	4	3.836	6.095
$Ca(Cr_{0.5}Ta_{0.5})O_3$	204.550	57.062	1	3.850	5.951
$Ca(Cr_{0.5}W_{0.5})O_3$	206.001	225.337	4	3.833	6.070
$Ca(Dy_{0.5}Nb_{0.5})O_3$	215.781	65.392	1	4.029	5.478
$Ca(Dy_{0.5}Ta_{0.5})O_3$	259.802	65.393	1	4.029	6.595
$Ca(Er_{0.5}Nb_{0.5})O_3$	218.161	64.756	1	4.016	5.592
$Ca(Er_{0.5}Ta_{0.5})O_3$	262.182	64.757	1	4.016	6.721
$Ca(Fe_{0.5}Mo_{0.5})O_3$	163.972	231.688	4	3.869	4.699
$Ca(Fe_{0.5}Nb_{0.5})O_3$	162.455	58.703	1	3.886	4.594
$Ca(Fe_{0.5}Sb_{0.5})O_3$	176.877	234.551	4	3.885	5.007
$Ca(Fe_{0.5}Ta_{0.5})O_3$	206.476	58.701	1	3.886	5.839
$Ca(Gd_{0.5}Nb_{0.5})O_3$	213.156	65.540	1	4.032	5.399

TABLE 9.2 *(cont.)*

Compound	Formula weight	Volume	Z	a', Å	Density
$A(B^{3+}_{0.5}B^{5+}_{0.5})O_3$ *(cont.)*					
$Ca(Gd_{0.5}Ta_{0.5})O_3$	257.177	65.541	1	4.032	6.514
$Ca(Ho_{0.5}Nb_{0.5})O_3$	216.996	64.912	1	4.019	5.549
$Ca(Ho_{0.5}Ta_{0.5})O_3$	261.017	65.237	1	4.026	6.642
$Ca(In_{0.5}Nb_{0.5})O_3$	134.531	62.222	1	3.963	3.589
$Ca(In_{0.5}Ta_{0.5})O_3$	178.552	62.381	1	3.966	4.751
$Ca(La_{0.5}Nb_{0.5})O_3$	203.986	67.373	1	4.069	5.026
$Ca(La_{0.5}Ta_{0.5})O_3$	248.007	67.372	1	4.069	6.111
$Ca(Mn_{0.5}Ta_{0.5})O_3$	206.021	58.851	1	3.890	5.811
$Ca(Nd_{0.5}Nb_{0.5})O_3$	206.651	66.368	1	4.049	5.169
$Ca(Nd_{0.5}Ta_{0.5})O_3$	250.672	66.371	1	4.049	6.270
$Ca(Ni_{0.5}W_{0.5})O_3$	209.358	230.769	4	3.864	6.024
$Ca(Pr_{0.5}Nb_{0.5})O_3$	204.985	66.699	1	4.055	5.102
$Ca(Pr_{0.5}Ta_{0.5})O_3$	249.006	66.701	1	4.056	6.197
$Ca(Sc_{0.5}Re_{0.5})O_3$	203.656	242.942	4	3.931	5.566
$Ca(Sm_{0.5}Nb_{0.5})O_3$	209.706	65.866	1	4.039	5.285
$Ca(Sm_{0.5}Ta_{0.5})O_3$	253.727	66.205	1	4.045	6.362
$Ca(Tb_{0.5}Nb_{0.5})O_3$	213.993	65.384	1	4.029	5.433
$Ca(Tb_{0.5}Ta_{0.5})O_3$	258.014	65.383	1	4.029	6.551
$Ca(Y_{0.5}Nb_{0.5})O_3$	178.984	65.232	1	4.026	4.555
$Ca(Y_{0.5}Ta_{0.5})O_3$	223.005	65.232	1	4.026	5.675
$Ca(Yb_{0.5}Nb_{0.5})O_3$	221.051	64.281	1	4.006	5.708
$Ca(Yb_{0.5}Ta_{0.5})O_3$	265.072	64.279	1	4.006	6.845
$Pb(Fe_{0.5}Nb_{0.5})O_3$	329.565	64.819	1	4.017	8.440
$Pb(Fe_{0.5}Ta_{0.5})O_3$	373.586	64.529	1	4.011	9.610
$Pb(In_{0.5}Nb_{0.5})O_3$	301.641	69.427	1	4.110	7.212
$Pb(Ho_{0.5}Nb_{0.5})O_3$	384.106	71.057	1	4.142	8.973
$Pb(Lu_{0.5}Nb_{0.5})O_3$	389.126	70.646	1	4.134	9.143
$Pb(Lu_{0.5}Ta_{0.5})O_3$	433.147	70.835	1	4.138	10.151
$Pb(Sc_{0.5}Nb_{0.5})O_3$	324.119	67.901	1	4.080	7.924
$Pb(Sc_{0.5}Ta_{0.5})O_3$	368.140	67.519	1	4.072	9.051
$Pb(Yb_{0.5}Nb_{0.5})O_3$	388.161	71.473	1	4.150	9.015
$Pb(Yb_{0.5}Ta_{0.5})O_3$	432.182	70.445	1	4.130	10.184
$Sr(Co_{0.5}Nb_{0.5})O_3$	211.538	60.698	1	3.930	5.785
$Sr(Co_{0.5}Sb_{0.5})O_3$	225.960	489.304	8	3.940	6.133
$Sr(Cr_{0.5}Mo_{0.5})O_3$	209.586	478.212	8	3.910	5.820
$Sr(Cr_{0.5}Nb_{0.5})O_3$	208.069	61.261	1	3.942	5.638
$Sr(Cr_{0.5}Os_{0.5})O_3$	256.716	481.890	8	3.920	7.075
$Sr(Cr_{0.5}Re_{0.5})O_3$	254.716	478.212	8	3.910	7.074
$Sr(Cr_{0.5}Sb_{0.5})O_3$	222.491	485.958	8	3.931	6.080
$Sr(Cr_{0.5}Ta_{0.5})O_3$	252.090	61.163	1	3.940	6.842
$Sr(Cr_{0.5}W_{0.5})O_3$	253.541	478.212	8	3.910	7.041
$Sr(Dy_{0.5}Ta_{0.5})O_3$	307.342	—	—	—	—

TABLE 9.2 *(cont.)*

Compound	Formula weight	Volume	Z	a', Å	Density
$A(B^{3+}_{0.5}B^{5+}_{0.5})O_3$ *(cont.)*					
$Sr(Er_{0.5}Ta_{0.5})O_3$	309.722	—	—	—	—
$Sr(Eu_{0.5}Ta_{0.5})O_3$	302.072	—	—	—	—
$Sr(Fe_{0.5}Mo_{0.5})O_3$	211.512	491.169	8	3.945	5.719
$Sr(Fe_{0.5}Nb_{0.5})O_3$	209.995	62.571	1	3.970	5.571
$Sr(Fe_{0.5}Sb_{0.5})O_3$	224.417	496.041	8	3.958	6.008
$Sr(Fe_{0.5}Ta_{0.5})O_3$	254.016	62.428	1	3.967	6.754
$Sr(Ga_{0.5}Nb_{0.5})O_3$	216.931	61.443	1	3.946	5.861
$Sr(Ga_{0.5}Os_{0.5})O_3$	265.578	478.212	8	3.910	7.375
$Sr(Ga_{0.5}Re_{0.5})O_3$	263.578	482.444	8	3.921	7.255
$Sr(Ga_{0.5}Sb_{0.5})O_3$	231.353	486.193	8	3.932	6.319
$Sr(Gd_{0.5}Ta_{0.5})O_3$	304.717	—	—	—	—
$Sr(Ho_{0.5}Ta_{0.5})O_3$	308.557	—	—	—	—
$Sr(In_{0.5}Nb_{0.5})O_3$	182.071	66.770	1	4.057	4.527
$Sr(In_{0.5}Os_{0.5})O_3$	230.718	523.607	8	4.030	5.852
$Sr(In_{0.5}Re_{0.5})O_3$	228.718	525.753	8	4.035	5.777
$Sr(In_{0.5}U_{0.5})O_3$	254.633	578.009	8	4.165	5.850
$Sr(La_{0.5}Ta_{0.5})O_3$	295.547	565.609	8	4.135	6.939
$Sr(Lu_{0.5}Ta_{0.5})O_3$	313.577	—	—	—	—
$Sr(Mn_{0.5}Mo_{0.5})O_3$	211.057	508.170	8	3.990	5.516
$Sr(Mn_{0.5}Sb_{0.5})O_3$	223.962	—	—	—	—
$Sr(Nd_{0.5}Ta_{0.5})O_3$	298.212	—	—	—	—
$Sr(Ni_{0.5}Sb_{0.5})O_3$	225.848	—	—	—	—
$Sr(Rh_{0.5}Sb_{0.5})O_3$	247.946	255.868	4	3.999	6.434
$Sr(Sc_{0.5}Os_{0.5})O_3$	253.196	515.850	8	4.010	6.518
$Sr(Sc_{0.5}Re_{0.5})O_3$	251.196	515.850	8	4.010	6.467
$Sr(Sm_{0.5}Ta_{0.5})O_3$	301.267	—	—	—	—
$Sr(Tm_{0.5}Ta_{0.5})O_3$	310.559	—	—	—	—
$Sr(Yb_{0.5}Ta_{0.5})O_3$	312.612	—	—	—	—
$A(B^{2+}_{0.5}B^{6+}_{0.5})O_3$					
$Ba(Ba_{0.5}Os_{0.5})O_3$	349.108	625.463	8	4.276	7.412
$Ba(Ba_{0.5}Re_{0.5})O_3$	347.108	623.271	8	4.271	7.396
$Ba(Ba_{0.5}U_{0.5})O_3$	373.023	702.595	8	4.445	7.051
$Ba(Ba_{0.5}W_{0.5})O_3$	345.933	636.056	8	4.300	7.223
$Ba(Ca_{0.5}Mo_{0.5})O_3$	253.348	583.229	8	4.177	5.769
$Ba(Ca_{0.5}Os_{0.5})O_3$	300.478	584.696	8	4.181	6.825
$Ba(Ca_{0.5}Re_{0.5})O_3$	298.478	583.439	8	4.178	6.794
$Ba(Ca_{0.5}Te_{0.5})O_3$	269.178	591.223	8	4.196	6.046
$Ba(Ca_{0.5}U_{0.5})O_3$	324.393	651.714	8	4.335	6.610
$Ba(Ca_{0.5}W_{0.5})O_3$	297.303	590.590	8	4.195	6.685
$Ba(Cd_{0.5}Os_{0.5})O_3$	336.638	576.969	8	4.162	7.748
$Ba(Cd_{0.5}Re_{0.5})O_3$	334.638	576.346	8	4.161	7.711

TABLE 9.2 *(cont.)*

Compound	Formula weight	Volume	Z	a', Å	Density
$A(B^{2+}_{0.5}B^{6+}_{0.5})O_3$ *(cont.)*					
$Ba(Cd_{0.5}U_{0.5})O_3$	360.553	321.487	4	4.316	7.447
$Ba(Co_{0.5}Mo_{0.5})O_3$	262.775	66.081	1	4.043	6.601
$Ba(Co_{0.5}Re_{0.5})O_3$	307.905	528.690	8	4.043	7.734
$Ba(Co_{0.5}U_{0.5})O_3$	333.820	587.217	8	4.187	7.549
$Ba(Co_{0.5}W_{0.5})O_3$	306.730	531.047	8	4.049	7.670
$Ba(Cr_{0.5}U_{0.5})O_3$	330.351	571.167	8	4.148	7.681
$Ba(Cu_{0.5}U_{0.5})O_3$	336.123	591.506	8	4.197	7.546
$Ba(Fe_{0.5}Re_{0.5})O_3$	306.362	521.660	8	4.025	7.799
$Ba(Fe_{0.5}U_{0.5})O_3$	332.277	574.271	8	4.156	7.684
$Ba(Fe_{0.5}W_{0.5})O_3$	305.187	537.963	8	4.066	7.534
$Ba(Mg_{0.5}Os_{0.5})O_3$	292.594	527.514	8	4.040	7.366
$Ba(Mg_{0.5}Re_{0.5})O_3$	290.594	527.906	8	4.041	7.310
$Ba(Mg_{0.5}Te_{0.5})O_3$	261.294	537.368	8	4.065	6.457
$Ba(Mg_{0.5}U_{0.5})O_3$	316.509	588.691	8	4.190	7.140
$Ba(Mg_{0.5}W_{0.5})O_3$	289.419	531.244	8	4.049	7.235
$Ba(Mn_{0.5}Re_{0.5})O_3$	305.907	547.343	8	4.090	7.422
$Ba(Mn_{0.5}U_{0.5})O_3$	331.822	618.470	8	4.260	7.125
$Ba(Ni_{0.5}Mo_{0.5})O_3$	262.663	65.086	1	4.022	6.699
$Ba(Ni_{0.5}Re_{0.5})O_3$	307.793	519.718	8	4.020	7.865
$Ba(Ni_{0.5}U_{0.5})O_3$	333.708	579.259	8	4.168	7.651
$Ba(Ni_{0.5}W_{0.5})O_3$	306.618	524.777	8	4.033	7.759
$Ba(Pb_{0.5}Mo_{0.5})O_3$	336.903	—	—	—	—
$Ba(Sr_{0.5}Os_{0.5})O_3$	324.248	619.686	8	4.263	6.949
$Ba(Sr_{0.5}Re_{0.5})O_3$	322.248	613.128	8	4.248	6.980
$Ba(Sr_{0.5}U_{0.5})O_3$	348.163	690.807	8	4.420	6.693
$Ba(Sr_{0.5}W_{0.5})O_3$	321.073	614.125	8	4.250	6.943
$Ba(Zn_{0.5}Os_{0.5})O_3$	313.123	530.457	8	4.047	7.839
$Ba(Zn_{0.5}Re_{0.5})O_3$	311.123	532.623	8	4.053	7.757
$Ba(Zn_{0.5}U_{0.5})O_3$	337.038	592.069	8	4.198	7.560
$Ba(Zn_{0.5}W_{0.5})O_3$	309.948	534.596	8	4.058	7.699
$Ca(Ca_{0.5}Os_{0.5})O_3$	203.218	261.552	4	4.029	5.159
$Ca(Ca_{0.5}Re_{0.5})O_3$	201.218	263.819	4	4.040	5.064
$Ca(Ca_{0.5}W_{0.5})O_3$	200.043	512.000	8	4.000	5.189
$Ca(Cd_{0.5}Re_{0.5})O_3$	237.378	260.017	4	4.021	6.062
$Ca(Co_{0.5}Os_{0.5})O_3$	212.645	235.445	4	3.890	5.997
$Ca(Co_{0.5}Re_{0.5})O_3$	210.645	234.899	4	3.887	5.954
$Ca(Fe_{0.5}Re_{0.5})O_3$	209.102	230.064	4	3.860	6.035
$Ca(Mg_{0.5}Re_{0.5})O_3$	193.334	236.743	4	3.897	5.423
$Ca(Mg_{0.5}W_{0.5})O_3$	192.159	456.533	8	3.850	5.590
$Ca(Mn_{0.5}Re_{0.5})O_3$	208.647	239.574	4	3.913	5.783
$Ca(Ni_{0.5}Re_{0.5})O_3$	210.533	231.998	4	3.871	6.026
$Ca(Sr_{0.5}W_{.5})O_3$	223.813	531.441	8	4.050	5.593

TABLE 9.2 *(cont.)*

Compound	Formula weight	Volume	Z	a', Å	Density
$A(B^{2+}_{0.5}B^{6+}_{0.5})O_3$ *(cont.)*					
$Pb(Ca_{0.5}W_{0.5})O_3$	367.153	—	—	—	—
$Pb(Cd_{0.5}W_{0.5})O_3$	403.313	70.620	1	4.133	9.480
$Pb(Co_{0.5}W_{0.5})O_3$	376.580	—	—	—	—
$Pb(Mg_{0.5}Te_{0.5})O_3$	331.144	510.082	8	3.995	8.621
$Pb(Mg_{0.5}W_{0.5})O_3$	359.269	64.000	1	4.000	9.319
$Sr(Ca_{0.5}Mo_{0.5})O_3$	203.628	—	—	—	—
$Sr(Ca_{0.5}Os_{0.5})O_3$	250.758	553.388	8	4.105	6.018
$Sr(Ca_{0.5}Re_{0.5})O_3$	248.758	276.644	4	4.105	5.971
$Sr(Ca_{0.5}U_{0.5})O_3$	274.673	304.017	4	4.236	5.999
$Sr(Ca_{0.5}W_{0.5})O_3$	247.583	551.368	8	4.100	5.963
$Sr(Cd_{0.5}Re_{0.5})O_3$	284.9182	271.657	4	4.080	6.964
$Sr(Cd_{0.5}U_{0.5})O_3$	310.833	300.066	4	4.217	6.878
$Sr(Co_{0.5}Mo_{0.5})O_3$	213.055	61.625	1	3.950	5.739
$Sr(Co_{0.5}Os_{0.5})O_3$	260.185	489.294	8	3.940	7.062
$Sr(Co_{0.5}Re_{0.5})O_3$	258.185	495.513	8	3.957	6.919
$Sr(Co_{0.5}U_{0.5})O_3$	284.100	549.353	8	4.095	6.868
$Sr(Co_{0.5}W_{0.5})O_3$	257.010	496.772	8	3.960	6.871
$Sr(Cr_{0.5}U_{0.5})O_3$	280.631	529.475	8	4.045	7.039
$Sr(Cu_{0.5}W_{0.5})O_3$	259.313	492.875	8	3.950	6.987
$Sr(Fe_{0.5}Os_{0.5})O_3$	258.642	483.737	8	3.925	7.100
$Sr(Fe_{0.5}Re_{0.5})O_3$	256.642	487.441	8	3.935	6.992
$Sr(Fe_{0.5}U_{0.5})O_3$	282.557	533.412	8	4.055	7.035
$Sr(Fe_{0.5}W_{0.5})O_3$	255.467	504.358	8	3.980	6.727
$Sr(Mg_{0.5}Mo_{0.5})O_3$	195.744	—	—	—	—
$Sr(Mg_{0.5}Os_{0.5})O_3$	242.874	489.294	8	3.940	6.592
$Sr(Mg_{0.5}Re_{0.5})O_3$	240.874	493.030	8	3.950	6.488
$Sr(Mg_{0.5}Te_{0.5})O_3$	211.574	500.566	8	3.970	5.613
$Sr(Mg_{0.5}U_{0.5})O_3$	266.789	549.353	8	4.095	6.449
$Sr(Mg_{0.5}W_{0.5})O_3$	239.699	493.039	8	3.950	6.456
$Sr(Mn_{0.5}Re_{0.5})O_3$	256.187	513.922	8	4.005	6.620
$Sr(Mn_{0.5}U_{0.5})O_3$	282.102	567.664	8	4.140	6.600
$Sr(Mn_{0.5}W_{0.5})O_3$	255.012	513.922	8	4.005	6.590
$Sr(Ni_{0.5}Mo_{0.5})O_3$	212.943	60.772	1	3.932	5.817
$Sr(Ni_{0.5}Re_{0.5})O_3$	258.073	488.050	8	3.937	7.022
$Sr(Ni_{0.5}U_{0.5})O_3$	283.988	541.343	8	4.075	6.967
$Sr(Ni_{0.5}W_{0.5})O_3$	256.898	488.677	8	3.938	6.981
$Sr(Pb_{0.5}Mo_{0.5})O_3$	287.183	—	—	—	—
$Sr(Sr_{0.5}Os_{0.5})O_3$	274.528	562.086	8	4.126	6.486
$Sr(Sr_{0.5}Re_{0.5})O_3$	272.528	575.019	8	4.158	6.294
$Sr(Sr_{0.5}U_{0.5})O_3$	298.443	323.356	4	4.324	6.128
$Sr(Sr_{0.5}W_{0.5})O_3$	271.353	551.368	8	4.100	6.536
$Sr(Zn_{0.5}Re_{0.5})O_3$	261.403	498.639	8	3.965	6.962

TABLE 9.2 *(cont.)*

Compound	Formula weight	Volume	Z	a', Å	Density
$A(B^{2+}_{0.5}B^{6+}_{0.5})O_3$ *(cont.)*					
$Sr(Zn_{0.5}W_{0.5})O_3$	260.228	502.438	8	3.975	6.878
$A(B^{1+}_{0.5}B^{7+}_{0.5})O_3$					
$Ba(Ag_{0.5}I_{0.5})O_3$	302.725	605.496	8	4.230	6.640
$Ba(Li_{0.5}Os_{0.5})O_3$	283.908	531.441	8	4.050	7.094
$Ba(Li_{0.5}Re_{0.5})O_3$	281.908	534.992	8	4.059	6.998
$Ba(Na_{0.5}I_{0.5})O_3$	260.285	578.010	8	4.165	5.980
$Ba(Na_{0.5}Os_{0.5})O_3$	291.933	567.869	8	4.140	6.827
$Ba(Na_{0.5}Re_{0.5})O_3$	289.933	570.961	8	4.148	6.744
$Ca(Li_{0.5}Os_{0.5})O_3$	186.648	480.049	8	3.915	5.163
$Ca(Li_{0.5}Re_{0.5})O_3$	184.648	480.049	8	3.915	5.108
$Sr(Li_{0.5}Os_{0.5})O_3$	234.188	485.588	8	3.930	6.405
$Sr(Li_{0.5}Re_{0.5})O_3$	232.188	487.443	8	3.935	6.326
$Sr(Na_{0.5}Os_{0.5})O_3$	242.213	537.368	8	4.065	5.986
$Sr(Na_{0.5}Re_{0.5})O_3$	240.213	537.368	8	4.065	5.936
$A^{3+}(B^{2+}_{0.5}B^{4+}_{0.5})O_3$					
$La(Co_{0.5}Ir_{0.5})O_3$	216.375	—	—	—	—
$La(Cu_{0.5}Ir_{0.5})O_3$	218.678	—	—	—	—
$La(Mg_{0.5}Ge_{0.5})O_3$	235.359	59.319	1	3.900	6 586
$La(Mg_{0.5}Ir_{0.5})O_3$	199.064	496.793	8	3.960	5.321
$La(Mg_{0.5}Nb_{0.5})O_3$	245.517	—	—	—	—
$La(Mg_{0.5}Ru_{0.5})O_3$	249.599	494.914	8	3.955	6.697
$La(Mg_{0.5}Ti_{0.5})O_3$	223.014	60.791	1	3.932	6.090
$La(Mn_{0.5}Ir_{0.5})O_3$	214.377	485.588	8	3.930	5.863
$La(Mn_{0.5}Ru_{0.5})O_3$	264.912	481.890	8	3.920	7.300
$La(Ni_{0.5}Ir_{0.5})O_3$	216.263	493.039	8	3.950	5.825
$La(Ni_{0.5}Ru_{0.5})O_3$	266.798	493.039	8	3.950	7.186
$La(Ni_{0.5}Ti_{0.5})O_3$	240.213	60.698	1	3.930	6.569
$La(Zn_{0.5}Ru_{0.5})O_3$	270.128	506.262	8	3.985	7.086
$Nd(Mg_{0.5}Ti_{0.5})O_3$	228.344	59.319	1	3.900	6.390
$A(B^{1+}_{0.25}B^{5+}_{0.75})O_3$					
$Ba(Na_{0.25}Ta_{0.75})O_3$	326.797	70.804	1	4.137	7.662
$Sr(Na_{0.25}Ta_{0.75})O_3$	277.077	66.676	1	4.055	6.898
$A(B^{3+}_{0.5}B^{4+}_{0.5})O_{2.75}$					
$Ba(In_{0.5}U_{0.5})O_{3.75}$	300.353	625.246	8	4.275	6.379
$A(B^{2+}_{0.5}B^{5+}_{0.5})O_{2.75}$					
$Ba(Ba_{0.5}Ta_{0.5})O_{3.75}$	340.482	656.235	8	4.345	6.890
$Ba(Fe_{0.5}Mo_{0.5})O_{3.75}$	257.232	527.514	8	4.040	6.476
$Sr(Sr_{0.5}Ta_{0.5})O_{3.75}$	265.902	580.094	8	4.170	6.087

There also is a small amount of mechanical property data available on perovskite solid solutions of the PZT type, $Pb(Zr_{1-x}Ti_x)O_3$.[5] The interest in these materials as piezoelectrics was discussed in Chapter 5. Tensile strength measurements on hot-pressed specimens of $Pb(Zr_{0.52}Ti_{0.48})O_3$, $Pb(Zr_{0.65}Ti_{0.35})O_3$, $Pb(Zr_{0.75}Ti_{0.25})O_3$ and $Pb(Zr_{0.95}Ti_{0.05})O_3$ containing 1 wt% Nb_2O_5 balanced with PbO gave values of 9700, 11,150, 12,550 and 14,600 psi respectively. The nominal grain size of these particles was 3 microns and the densities of the specimens were at least 95% theoretical. The moduli of the hot-pressed specimens of $Pb(Zr_{0.65}Ti_{0.35})O_3$, $Pb(Zr_{0.75}Ti_{0.25})O_3$ and $Pb(Zr_{0.95}Ti_{0.05})O_3$ described above were found to be 14×10^6 psi for the former two phases and 16×10^6 for the latter phase.

References

1. W. R. EPPERLY et al., Semiannual Report No. 7 (1 Jan. 1965–30 June 1965).
2. P. G. DICKENS and M. S. WHITTINGHAM, Trans. Faraday Soc. **61**, 1226 (1967).
3. G. PARRAVANO, J. Chem. Phys. **20**, 342 (1952).
4. J. R. TINKLEPAUGH, presented at the 12th Sygamore Army Mat. Res. Conf. (24–27 Aug. 1965).
5. R. H. DUNGAN, Development Report SC–DR–66–593, Sandia Corp. (Nov. 1966).

CHAPTER 10

PREPARATION OF PEROVSKITE-TYPE OXIDES

10.1. Powders

Many of the perovskite-type oxides can be prepared by a high-temperature solid-state reaction between binary oxide powders which are stable in air. However, it is often advantageous to use carbonates, nitrates, etc., instead of the oxides, if they can be obtained in smaller particle sizes to insure quicker reactions or if they are much purer. In a typical preparation of barium titanate powder, barium carbonate and titanium oxide are weighed out in equal molar quantities, mixed thoroughly and fired at 1000°C for 24 hr in zirconium silicate boats or in platinum crucibles. However, the material often will react with the platinum or impurities in the boats as evidenced by coloration of the sample. Therefore, if extremely pure material is desired it is advantageous to fire the mixture on top of a barium titanate compact. In addition, since the oxides often can not be entirely reacted by a single firing, repeated regrinding and reheating is required.

Some of the perovskite-type compounds can only be prepared using special techniques. These are discussed below.

Compounds containing lead in the A position of the perovskite structure are often difficult to prepare because of the volatility of lead oxide. Sometimes this problem can be alleviated by heating the reactants in a lead oxide atmosphere, by using an excess of lead oxide in the reaction mixture, or by heating the reactants at a relatively low temperature to allow them to combine before the final firing. When lead is the A ion and pentavalent ions are used as the B ions, compounds with the pyrochlore structure very often are formed early in the reaction sequence and are difficult to react further even with repeated mixing and reheating of the product.

Fresia et al.[1] found that in the preparation of compounds with the general formula $A(B^{2+}_{0.5}W_{0.5})O_3$, where $A = Sr$, Ba, $B^{2+} = Fe$, Co, Ni and Zn, alkaline earth tungstates always could be detected in the final products. Similar results have been observed for molybdenum compounds. Regrinding and refiring of the samples often helps to reduce the amount of tungstate or molybdate present, but each compound has to be treated as a separate case for which the best firing temperature and firing time must be determined.

In preparing perovskite-type compounds containing divalent iron and cobalt, the valence state is retained by heating the sample in an evacuated sealed silica capsule or in a non-oxidizing atmosphere. Divalent Fe or Cr ions can be obtained in a compound by mixing equal amounts of metallic Fe and Fe_2O_3 or Cr and Cr_2O_3, respectively, with the other oxide constituents. For example, Galasso et al.[2] prepared $Ba(Fe^{2+}_{0.33}Ta_{0.67})O_3$ by mixing the reactants according to the equation:

$$BaO + \tfrac{1}{9}Fe_2O_3 + \tfrac{1}{9}Fe + \tfrac{1}{3}Ta_2O_5 \longrightarrow Ba(Fe_{0.33}Ta_{0.67})O_3,$$

compacting the sample and firing it at 1000°C for 24 hr in an evacuated sealed silica capsule.

Sleight and Ward[3] in forming compounds of the $A(B^{2+}_{0.5}U^{6+}_{0.5})O_3$ type found it advantageous to use $UO_2(NO_3)_2 \cdot 6H_2O$ as a source of hexavalent uranium. Compounds containing pentavalent uranium were prepared by heating UO_2 and UO_3 in equal proportions with the other oxide reactants. The UO_3 was obtained by heating $UO_2(NO_3)_2 \cdot 6H_2O$ at 400°C in air and the UO_2 was prepared by heating UO_3 at 1000°C in hydrogen.

Sleight et al.[4] also produced other complex perovskite compounds such as $Sr(Na_{0.5}Re_{0.5})O_3$ and $Sr(Na_{0.5}Os_{0.5})O_3$ which contained heptavalent rhenium and osmium by heating the metal with sodium carbonate and strontium oxide in air. A similar procedure was used to prepare compounds with hexavalent osmium as one of the B ions. The reaction is given as:

$$AO + \tfrac{1}{2}BO + \tfrac{1}{2}Os + \tfrac{3}{4}O_2 \longrightarrow A(B_{0.5}Os_{0.5})O_3$$

where A = Sr or Ca, and B = a divalent metal ion.

A mixture of (ReO$_3$ + Re) was used in the preparation of compounds containing pentavalent rhenium with other trivalent ions in the B position.

Patterson et al.[5] found that they could prepare $A(B^{3+}_{0.5}W^{5+}_{0.5})O_3$ and $A(B_{0.5}Mo_{0.5})O_3$ type compounds by using a mixture of the metal trioxides and the metals to obtain pentavalent tungsten and pentavalent molybdenum respectively.

Ridgley and Ward[6] prepared the strontium–niobium bronzes which contained some niobium ions in the tetravalent state. These phases adopted the perovskite structure when x varied between 0.7 and 0.95 in Sr_xNbO_3. Two procedures were used to prepare these bronzes. The first involved the following reaction:

$$(0.5+x)SrO + 0.4xNb + (0.5-0.2x)Nb_2O_5 \longrightarrow Sr(0.5+x)NbO_3$$

where Nb metal was used to reduce part of the pentavalent niobium and the second required that NbO$_2$ be formed first and reacted with SrO and Nb$_2$O$_5$. The NbO$_2$ was prepared by reducing Nb$_2$O$_5$ with hydrogen at 1200°C for 36 hr.

Other methods for producing compounds containing elements in unusual oxidization states have been reported by McCarroll et al.[7] Randall and Ward[8] and Kestigian et al.[9] McCarroll et al. prepared CaMoO$_3$ using a mixture of molybdenum metal and MoO$_3$ as a source of tetravalent molybdenum, Randall and Ward prepared SrRuO$_3$ by heating strontium oxide and ruthenium metal in air, and Kestigian et al., prepared LaVO$_3$ by mixing and heating La$_2$O$_3$ and V$_2$O$_3$ together in vacuum. The V$_2$O$_3$ was obtained by heating vanadium pentoxide in hydrogen at 800° for 14 hr.

Many of the ions in binary oxides which are unstable become stabilized in the perovskite structure. For example, BaFe^{4+}O$_3$ can be prepared although heating Fe$_2$O$_3$ in oxygen will not produce Fe^{4+}. As another example, the addition of a rare earth oxide or Y$_2$O$_3$ to barium titanate produces some Ti^{3+} when the mixture is heated to 1000°C in air. This material is extremely stable, even though heating Ti$_2$O$_3$ in air would result in the formation of TiO_2. This factor is used to advantage in the preparation of many of the perovskite type compounds.

The preparation of the magnetic perovskites $BiMnO_3$ and $BiCrO_3$ created an interesting new area of research on perovskite type compounds.[10] These compounds were formed by heating the oxides at 700°C under a pressure of 40 kbar, and then quenching them. In both cases, they formed compounds which had distorted perovskite structures with triclinic unit cells.

10.2. THIN FILMS

The need for special elements in microcircuitry has caused considerable interest in the formation of thin films of dielectric materials. The perovskite-type compounds which have high dielectric constants are most attractive for this purpose. However, the problems of forming binary oxides in thin film form have not been entirely solved, thus, researchers have been reluctant to attempt the preparation of thin films of more complicated materials. Of the studies which have been conducted on perovskites, most of them have been on $BaTiO_3$ films. Films 7.5 μ thick have been formed by a special slip method[11] but thinner films could not be prepared. In addition, thin single crystal films have been prepared by first growing crystals from solution[12] and then etching in hot phosphoric acid.[13-14] Films also have been obtained by spreading small amounts of molten $BaTiO_3$ on a platinum sheet.[15]

In 1955 Feldman prepared ferroelectric films of $BaTiO_3$ approximately 1 to 2 μ thick by vapor deposition.[16] In this process, the $BaTiO_3$ powder was mixed with alcohol, placed on a tungsten-wire filament and vacuum evaporated onto a platinum substrate. The BaO evaporated first and the TiO_2 later, but after firing the film at 1000–1100°C the barium titanate was again formed. X-ray diffraction studies indicated that the films consisted of mainly $BaTiO_3$ with traces of BaO_2 and TiO_2 and with minor amounts of $BaTi_4O_9$, $BaTi_3O_7$ and Ba_2TiO_4. The samples were prepared for property measurements by placing a gold dot on the surface to act as an electrode. The films were found to be ferroelectric and had dielectric constants as high as 270. This, of course, is one of the simpler techniques of forming barium titanate films, but is not satisfactory for obtaining pure materials.

Using a more elaborate method, Green[17] deposited alternating layers of BaO and TiO_2 by successive evaporation from several tungsten coils and then heating the films in air at 1150°C. Frankl et al.[18] used two electron beams to evaporate TiO_2 and BaO and Moll[19] evaporated barium titanate in an electric field to obtain single-crystal titanate films; however, Roder's studies[20] left some doubt as to the reliability of Moll's process.

The best technique for producing thin films of barium titanate was used by Müller et al.[21] who prepared thin films with thicknesses of the order of 1 μ by vacuum evaporating grain by grain of powder. In addition, solid solutions of $BaTiO_3$ with $SrTiO_3$ and $BaSnO_3$ were prepared in thin film form by the same technique. The samples were formed into pellets crushed and sieved to form grains of 100/200-mesh size. The grains then were delivered by a V-shaped niobium trough and were moved by taps from a gear wheel transmitted through a rod (see Fig. 10.1). The grains were dropped onto an iridium boat where a liquid pool of the material was maintained and from which the material was evaporated onto a substrate held at approximately 500°C. The boat temperature was held 2300°C for the $BaTiO_3$ evaporation, slightly lower for those containing some $BaSnO_3$. When the substrate was held at room temperature, amorphous films were formed, but using a substrate temperature of 500°C resulted in the formation of crystalline films. The crystalline films of $BaTiO_3$ were found to have a dielectric constant of 400 to 700. Substitution of strontium or tin into the barium titanate before evaporation resulted in films with lower dielectric constants, 250–300 for $Ba(Ti_{0.9}Sn_{0.1})O_3$ and 200–400 for $Sr_{0.73}Ba_{0.27}TiO_3$. The dissipation factor at 1 kc/s for 0.2 μ films was larger for crystalline films of barium titanate than for the amorphous films and the breakdown strength of the amorphous films exceeded 1.5×10 V/cm while that of the crystalline films was 2 to 3 times lower. However, the dielectric constant of the amorphous films was only from 13–20.

It should be noted that the films also were prepared using a flash evaporation technique where the grain is evaporated from the boat before the next one arrives, but the technique of using a molten pool was preferred by Müller because it reduced contamination from the boat and permitted the use of lower heating temperatures.

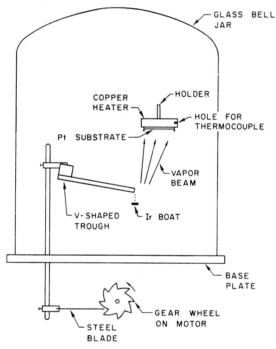

FIG. 10.1. Diagrammatic view of evaporation system (after Müller et al.[21]).

Müller et al. also deposited thin films of $SrTiO_3$, $CaTiO_3$, $BaSnO_3$, $SrSnO_3$, $BaCeO_3$ and $NaNbO_3$ using the grain by grain evaporation technique.[22] A cleaved LiF substrate was rigidly fastened to a copper block and heated to 700°C. The results of these experiments indicate that the technique may be feasible for producing single crystals of these materials.

Attempts also have been made to produce thin films of $PbTiO_3$ with tolerances of $\pm 0.1\ \mu$ by generating a plasma of the bulk material in a vacuum chamber and forming the thin films on a low temperature substrate.[23] Thin films have been formed with capacitances of the order 50 $\mu F/in^2$ indicating a dielectric constant of over 100. The breakdown voltage of the thinnest films has been found to be 10 volts.

10.3. SINGLE CRYSTALS

Of the large number of compounds with the perovskite structure, those with ferroelectric and magnetic properties and those with potential application as laser host materials have received the most attention from single-crystal researchers.

$KNbO_3$

Single crystals of potassium niobate, $KNbO_3$ and $BaTiO_3$ have been studied most extensively. Matthias and Remeika[24]

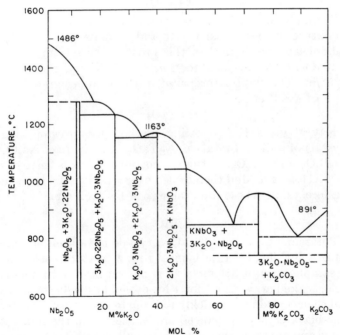

FIG. 10.2. Phase diagram of the K_2CO_3–Nb_2O_5 system (after A. Reisman and F. Holtzberg, J. Am. Chem. Soc. 77, 2117 (1955)).

have reported the growth of $KNbO_3$ single crystals using KCl or KF as a flux. In later studies, Shirane et al.[25] and Triebwasser[26] also grew $KNbO_3$ from a flux. The phase diagram is shown in Fig. 10.2. Pulvari[27] evaluated the methods of these previous workers and tried additional fluxes and flux combinations to obtain pure crystals. The fluxes used were K_2CO_3, KCl, NaCl, KF, $CaCl_2$, KBO_2, K_2SiO_3 with K_2CO_3 being selected as the most satisfactory one for growing ferroelectric crystals. The purest crystals were grown with 5–10% excess of K_2CO_3 and a soak time in the 1080–1100°C temperature range before slowly cooling the melt. Repeated recrystallization resulted in clearer and less colored crystals. The problem with the flux technique in general is that the crucible is often attacked and the crucible material and solvent enters in the growth process.

$NaNbO_3$

Sodium niobate, $NaNbO_3$, crystals have been grown from a mixture of sodium carbonate and niobium pentoxide in a sodium fluoride flux.[28] In this process, the mixture is preheated to 1000°C, soaked for 2 hr at 1350°C and then cooled at 5°/hr. The crystals prepared in this manner grew in the form of small cubes.

$NaTaO_3$

Kay[29] gives the following method for growing single crystals of sodium tantalate, $NaTaO_3$. A mixture of Na_2CO_3, $Na_2B_4O_7$ and Ta_2O_5 in the proportions 7:1:4 is heated at 1200°C for 12 hr and then cooled over a period of 6 hr resulting in crystals whose dimensions are $1 \times 2 \times 2$ mm.

$KTaO_3$

Potassium tantalate, $KTaO_3$, crystals have been prepared using a KF flux.[30] In a typical run, a mixture with a flux to sample ratio of 5:1 mole % is melted in a platinum crucible, soaked at > 1300° for 4 hr, cooled at 30°/hr to 900°C, and then cooled more quickly to room temperature. The dark blue-black crystals produced are leached from the flux with water and are then used as seeds to pull clear $KTaO_3$ crystals from a melt. These blue crystals are not adequate for ferroelectric applications since they are highly conducting.

In the process used by Wemple[30] to pull crystals of $KTaO_3$, a typical charge of 70 g of Ta_2O_5, 42.10 g of K_2CO_3, 24 mg of MnO_2 and 12 mg SnO_2 are mixed, placed in a 100-ml platinum crucible, and set into a vertical furnace. An oxygen atmosphere is maintained in the furnace (see phase diagram, Fig. 10.3). The mixture is slowly heated to a temperature

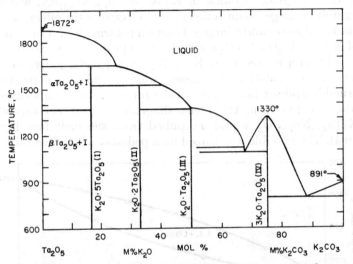

FIG. 10.3. Phase diagram of the K_2CO_3–Ta_2O_5 system (after A. Reisman, F. Holtzberg, M. Berkenblit and M. Berry, J. Am. Chem. Soc. 78, 4514 (1956)).

10–20°C above the liquidus over a 12–15-hr period and soaked for 4–6-hr. The $KTaO_3$ seed then is lowered to within 0.5 cm from the melt surface, the temperature of the melt is raised 5°C above the liquidus temperature, T_L, and slowly cooled at the rate of 3.5°C/hr. When the melt passed through a temperature 2–3°C above T_L, the seed is lowered to touch the melt and lifted 1–2 mm, pulling a small meniscus. The seed drive motor then is set to rotate at 60 rev/min with reversal every 30 sec. During the growth period the cooling rate is maintained at 3.5°C/hr and at various times the seed is lifted 1–2 mm. When the crystal is at the desired size, it is lifted above the melt surface but still kept in the furnace, and the cooling rate is changed to 25–30°C/hr until room temperature is attained. The color of the crystals grown in this manner changed from a bright green to colorless at room temperature and the crystals weighed from 4–10 g.

KTN

A similar procedure was used by Wemple[30] and Bonner et al.[31] to grow crystals of KTN, $K(Ta_{0.63}Nb_{0.37})O_3$, which exhibit a large room-temperature electro-optic effect, low electrical losses and a large saturation polarization. A mixture of K_2CO_3, Ta_2O_5, Nb_2O_5 and SnO_2 in appropriate proportions to obtain a composition $K_{1+x}(Ta_{0.29}Nb_{0.71})O_3Sn_{0.001}$, where x depends on volatilization losses, were placed in a platinum crucible and set in a vertical tube furnace. The composition was selected from the phase diagram so that crystals of $K(Ta_{0.63}Nb_{0.37})O_3$ would be pulled from the melt (see Fig. 10.4). The crucible was placed on a pedestal which was rotat-

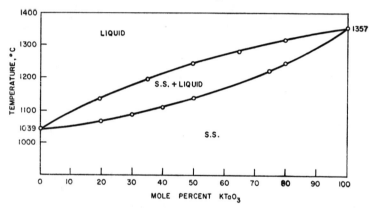

FIG. 10.4. Phase diagram of the $KNbO_3$–$KTaO_3$ system (after A. Reisman, S. Triebwasser and F. Holtzberg, *J. Am. Chem. Soc.* **77**, 4228 (1955)).

ed at 60 rev/min with reversals every 30 sec. The top of the muffle which contained the crucible was closed by a split ring with a hole to accommodate the rod that held the seed. The furnace was held at approximately 1225°C, oxygen flow was maintained up through the muffle, and the bottom of the crucible was kept 50°C warmer than the top of the melt. The melt was then cooled 0.1°C/hr while the seed was rotated at the surface. The seed was allowed to grow laterally from 2 to 3 days, and then the growing crystal was lifted at $\frac{1}{4}$ in. per day until it was cool enough to remove from the furnace. It is difficult in this process to produce a crystal with a uniform composition throughout.

$BaTiO_3$

Because of its ferroelectric properties, $BaTiO_3$ single crystals have been studied extensively. Crystals of $BaTiO_3$ have been grown from fluxes, by the Czochralski technique, the Verneuil technique and also using zone melting methods.

Some of the fluxes used to grow $BaTiO_3$ include $BaCl_2$,[32] KF[33] and BaF_2.[34] Linares obtained small blue butterfly twins (shaped like butterfly wings) and cubes using BaF_2 as a flux.

Potassium fluoride is a more popular flux than the abovementioned fluxes for growing $BaTiO_3$ single crystals. In this technique, a mixture of 10 mole % $BaTiO_3$ and 90 mole % KF is soaked at temperatures of 980 to 1200°C for 4 hr followed by cooling rates of 10°C/hr, resulting in butterfly crystals being formed. DeVries[35] conducted detailed studies on these butterfly twin crystals and concluded that they were probably formed when (111) twinning of the (100) habit takes place.

Single crystals of $BaTiO_3$ also have been grown by pulling from a $BaO-TiO_2$ melt by von Hippel *et al.* The phase diagram is shown in Fig. 10.5. A detailed schematic of the crystal growth furnace is shown in Fig. 10.6. The mixture used was made up of 67 mole % TiO_2 and 33 mole % BaO. The mixture was soaked at 1420°C, the temperature was lowered to 1396° and the seed then was immersed about 1 mm below the surface with air flowing through the seed rod. After 30 min the melt was cooled 5°C/hr for 1 hr. The crystal then was pulled at 0.25 mm/hr and the melt was cooled at 2 to 3°C/hr. The seed was rotated at 60 rev/min and reversed at 30-sec intervals. Then, the crystal was removed from the melt at 1335°C and annealed to room temperature. Single crystals of $BaTiO_3$ up to 2 cm diameter by 1 cm long have been grown by this method.

Attempts also were made by von Hippel et al.[36] to grow single crystals of $BaTiO_3$ by the flame-fusion technique. For this technique powders with small particle size but with good flow characteristics are necessary. In order to produce the powder, a solution of titanium tetrachloride, prepared by dripping 1.50 moles of $TiCl_4$ into 500 ml of water below 20°C, was added to a solution of oxalic acid, 5 moles $(COOH)_2$ in 1320 ml of water and held at 20°C. A solution of barium chloride at 70°C, 1.60 moles $BaCl_2 \cdot 2H_2O$ dissolved in 900 ml

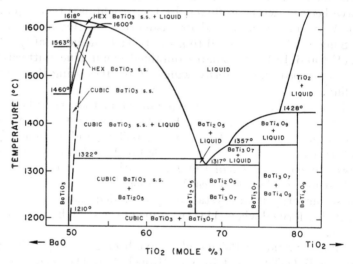

FIG. 10.5. Phase diagram of the $BaO-TiO_2$ system (after D. E. Rase and R. Roy, J. Am. Ceram. Soc. 38, 110 (1955)).

H_2O, was added to the mixture with rapid stirring. After 5 hr of stirring, barium titanyl oxalate was filtered, washed and ignited for 2 hr at 1000°C. Powder was passed through a sieve, and was used in the flame-fusion apparatus (see Fig. 10.7). Small crystals of $BaTiO_3$ were obtained, but they were not of optical quality.

Brown and Toat[37] found that they could grow single crystals 2.5 cm long by 0.32 cm diameter of $BaTiO_3$ which contain 1.5%, $SrTiO_3$ by a floating zone process. The heating source was a ring burner closely surrounding the molten zone.

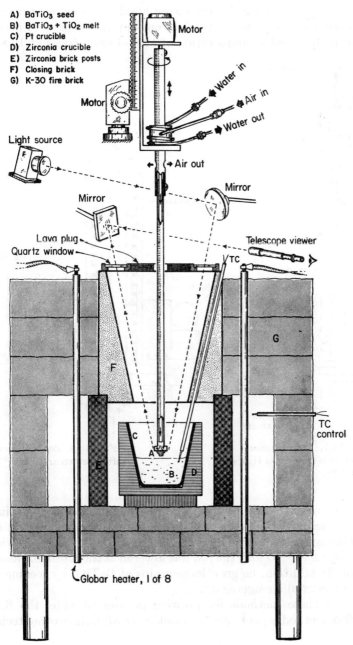

FIG. 10.6. Crystal-pulling furnace (after von Hippel et al.[36]).

Pressed bars of the material were fused to a seed crystal and were withdrawn at a rate of 3.5 cm/hr into an auxiliary heater maintained at temperature of 1300–1500°C (see Fig. 10.8).

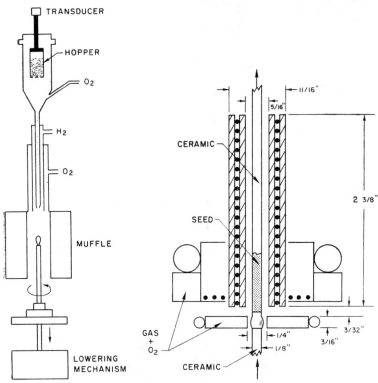

Fig. 10.7. Flame-fusion apparatus (after von Hippel et al.[36]).

Fig. 10.8. Zone melter (after Brown and Toat[37]).

In an interesting study, DeVries[38] obtained large grains of barium titanate in a polycrystalline rod by lowering it through a steep gradient. The center of the furnace coil was only at 1300°C and the rod was lowered at the rate of 0.6 mm/hr. In addition, he grew large grains of $BaTiO_3$ by seeding a polycrystalline aggregate.

Of these methods for growing barium titanate, the KF flux method is probably the most successful, since other techniques such as pulling from the melt and the Verneuil method involve relatively rapid and uneven cooling of the crystal through a phase transition which causes the crystal to crack, but there are problems with all of the crystal-growing techniques. The flux technique and the Czochralski method both require the use of a crucible which may dissolve and be introduced into the crystal. In the Verneuil technique special powders must be prepared and the resulting crystals are highly strained. The floating-zone process produces crystals with less strain than the Verneuil technique, but with more defects than crystals grown by flux and Czochralski techniques.

$CaTiO_3$

Single crystals of $CaTiO_3$ have been obtained using $CaCl_2$, $BaCl_2$, $CaCl_2-BaCl_2$ and $Na_2CO_3-K_2CO_3$ fluxes. The chloride solutions were soaked at 1150°C and the carbonates at 1000°C for 30–40 hr. Cooling rates used varied from 50°C to 100°C/hr. The crystals produced were of the order of 1 mm.[39]

Larger crystals of $CaTiO_3$, 25 mm long and 12 mm in diameter, have been grown by Merker using the flame-fusion technique.[40] The feed material was prepared by mixing solutions of titanium tetrachloride, calcium chloride and oxalic acid in molar proportions of 1.0/1.4/4.0. Then the calcium titanyl oxalate formed was transformed to $CaTiO_3$ by heating to 1000°C. After passing the $CaTiO_3$ through a 100-mesh sieve, the powder was placed in the flame-fusion apparatus and the crystal was grown. Initially, the boules fractured after growth, but this was overcome by annealing and slowly cooling the crystal.

$SrTiO_3$

Strontium titanate crystals also have been prepared by Merker.[41] The feed material was formed by nearly the same process reported for preparing $BaTiO_3$ boule powder. A solution of oxalic acid was added to dilute titanium tetrachloride, followed by the addition of a solution of strontium chloride. The solution temperature was held at 70°C under agitation. After aging, the crystal salt was filtered, heated at 1000°C, sieved and used. The crystal as prepared had an opaque black appearance; however, a colorless transparent crystal was obtained by annealing the crystal in air.

$PbTiO_3$

Single-crystal growth of $PbTiO_3$ is of interest because $PbTiO_3$ is a high-temperature ferroelectric. Rogers[42] grew clear crystals $5 \times 3 \times 0.2$ mm of $PbTiO_3$ using a Bridgman–Stockbarger method with excess PbO. Nomura and Sawada also obtained similar results using a $PbCl_2$ flux.[43]

$CdTiO_3$

Single crystals of $CdTiO_3$, $0.12 \times 0.06 \cdot 0.06$ mm, have been grown from a NaCl flux.[44]

$PbZrO_3$

Pulvari[45] grew crystals of $PbZrO_3$ using the techniques reported by Jona et al.,[46] which involved placing a mixture of 2.4 g of PbF_2 and 6.9 g of $PbZrO_3$ in a covered platinum crucible, heated it at 1250°C for 1 hr and cooling at a rate of 50°/hr. The crystals grown were in the shape of cubes, 3 mm on edge.

(R.E.) BO_3

A large number of small perovskite crystals with Y, La, Pr, Nd, Eu, Sm and Gd in the A position and Al, Sc, Cr, Fe, Co and Ga in the B position were grown by Remeika.[47] A constituent oxide to lead monoxide ratio of 1:6 by weight was mixed and a platinum crucible was used to hold the mixture. The mixture was maintained at a temperature of 1300°C for a short period, except for compounds containing Al^{3+} or Sc^{3+} where the soaking time used was 4 hr. The temperature was then reduced to 850°C at a rate of 30°/hr and the crystals were leached from the mixture and crucible with hot dilute nitric acid.

$LaAlO_3$

Because of the longer fluorescence lifetime of Cr^{3+} in $LaAlO_3$ compared to Cr^{3+} in any other laser host material, the crystal is considered a good candidate for a high power-pulsed laser. However, the phase transition $LaAlO_3$ exhibits at 435°C presents a problem for growing it as a single crystal by many of the popular techniques.

Single crystals of pale yellow $LaAlO_3$ measuring $\frac{1}{2} \times \frac{1}{2} \times \frac{1}{4}$ in. have been grown by Airtron[48] in 250-ml crucibles from Bi_2O_3–B_2O_3 flux. A Bi_2O_3 flux containing 18.93 mole % B_2O_3 and 81.7 mole % Bi_2O_3 was mixed in proportions with the other oxides so that there were 10.97 and 10.77 moles of La_2O_3 and Al_2O_3 respectively. The platinum crucible was placed in a furnace, heated for 16 hr at 1340°C while rotating it and cooling it slowly to 960°C. The flux was poured off and the crystals washed with dilute HNO_3. All of the crystals were in the form of rectangular parallelepipeds. Airtron also has produced growth on a seed using a hydrothermal synthesis process where $LaAlO_3$ powder was used for a nutrient. The K_2CO_3 concentration used was 7 molal, the pressure was 20,000 psi, and the growth temperature was approximately 500°C.

Single crystals of $LaAlO_3$ also have been prepared by pulling from the melt.[49] The crucible was charged with La_2O_3 and Al_2O_3 powder in equal quantities. The melting point of $LaAlO_3$ was found to be 2075 − 2080°C. From this melt, crystals up to 43 g in weight were grown and a total of five different boule axis orientations were obtained.

$GdAlO_3$

In a study of laser host materials, Mazelsky et al.[50] prepared gadolinum aluminum oxide, $GdAlO_3$, single crystals from a melt. The seed was inserted in the melt and pulled at a temperature of 2030°C while rotating the rod holding the seed at 65 rev/min. Using a pull rate of 6 − 7 mm/hr crystals $\frac{1}{2}$ in. in diameter and 1 in. long were obtained. The crystals, however, were not of good optical quality.

Na_xWO_3

The sodium tungsten bronzes, Na_xWO_3, were prepared by Straumanis.[51] The method is based on the reaction

$$3xNa_2WO_4 + (6-4x)WO_3 + xW \longleftrightarrow 6\,Na_xWO_3.$$

To prepare orange to yellow bronzes, perovskite type, in which x is 0.6 to 0.9, it was necessary to use mole ratios of reactants 4:2:1 to 9:2:1. The mixture was heated to 850°C in an inert atmosphere and then slowly cooled. The crystals were recovered by leaching the solid in boiling water, sodium hydroxide solution and then HF.

$Pb(B^{2+}_{0.33}B_{0.67})O_3$

Bokov and Myl'nikova[52] prepared single crystals of the ferroelectric compounds $Pb(Ni_{0.33}Ta_{0.67})O_3$, $Pb(Mg_{0.33}Ta_{0.67})O_3$ $Pb(Co_{0.33}Nb_{0.67})O_3$, $Pb(Co_{0.33}Ta_{0.67})O_3$, $Pb(Zn_{0.33}Nb_{0.67})O_3$ using a lead oxide flux. A mixture of 60–80 mole % PbO and reagent grade oxides in proper proportions was placed in a platinum crucible and heated to 1200–1300°C. The melt was cooled at a rate of between 30–100°C/hr to a temperature of 800°C and then cooled more rapidly to room temperature. Crystals were separated by boiling in 20% HNO_3 except for $Pb(Zn_{0.33}Nb_{0.67})O_3$ crystals which were washed in acetic acid. The crystals were in the form of imperfect cubes, 1 – 2 mm on edge.

Later, Bokov et al.[53] grew $Pb(Co_{0.5}W_{0.5})O_3$ single crystals using a similar technique. Cobalt carbonate, H_2WO_4 and PbO were mixed in amounts corresponding to 10 mole % CoO, 20 – 30 mole % WO_3, and 70 – 80% PbO. The soaking temperature was 1200°C and the cooling rate to 800°C was 5°C/hr. The crystals formed were small cubes.

$Ba(B^{2+}_{0.33}Ta_{0.67})O_3$

Single crystals of $Ba(B^{2+}_{0.33}Ta_{0.67})O_3$ type compounds[54] where B^{2+} is Ca, Mg, Zn or Ni were grown by Galasso and Pinto using a BaF_2 flux. The details of the process for each compound and results are presented in Table 10.1.

$Pb(B^{3+}_{0.5}B^{5+}_{0.5})O_3$

Ferroelectric crystals of $Pb(B^{3+}_{0.5}B^{5+}_{0.5})O_3$ type compounds, where B^{3+} is Sc or Fe and B^{5+} is Nb or Ta, were prepared by Galasso and Darby[55] using PbO and $PbO-PbF_2$ fluxes. The conditions of growth and the results are as in Table 10.2.

TABLE 10.1. *Crystal Growth Data for* $Ba(B_{0.33}^{2+}Ta_{0.67})O_3$-*type Compounds (after Galasso and Pinto*[54]*)*

Compound	Max. temp. (°C)	Soaking time	Sample wt(g)	Flux wt(g)	Cooling rate (°C/hr)	Crystal size (mm)	Color
$Ba(Ca_{0.33}Ta_{0.67})O_3$	1365	1.0 hr	6.4	24.0	13	1.0	yellow
$Ba(Mg_{0.33}Ta_{0.67})O_3$	1400	8.5 hr	6.4	43.0	10	2.0	yellow
$Ba(Ni_{0.33}Ta_{0.67})O_3$	1360	0.5 hr	6.3	18.0	40	0.5	green
$Ba(Zn_{0.33}Ta_{0.67})O_3$	1380	2.0 hr	11.2	65.0	10	1.5	red

TABLE 10.2. *Crystal Growth Data for* $Pb(B_{0.5}^{3+}B_{0.5}^{5+})O_3$-*type Compounds (after Galasso and Darby*[55]*)*

Compound	Flux	Flux (wt%)	Temp range (°C)	Cooling rate (°C/hr)	Crystal size (mm) on edge
$Pb(Fe_{0.5}Nb_{0.5})O_3$	PbO	64	1230–800	5	1
$Pb(Fe_{0.5}Ta_{0.5})O_3$	PbO	54	1230–800	5	1
$Pb(Sc_{0.5}Nb_{0.5})O_3$	PbO	86	1150–900	30	1
$Pb(Sc_{0.5}Ta_{0.5})O_3$	PbO–PbF$_2$	42.5	1325–1025	25	1

$$Ba(B_{0.5}^{3+}Ta_{0.5})O_3$$

Galasso et al. studied $Ba(B_{0.5}^{3+}Ta_{0.5})O_3$-type materials for laser application.[56] Single crystals of these compounds with B^{3+} = La, Gd, Lu, Sc and Y were grown using BaF$_2$ flux, but more satisfactory results were obtained with B_2O_3 as the flux. Crystals of $Ba(Y_{0.5}Ta_{0.5})O_3$ up to 0.5 cm on edge were formed by mixing 566.3 g BaCO$_3$, 164.1 g of Ta$_2$O$_5$, 83.9 g Y$_2$O$_3$ and 112 g B$_2$O$_3$, placing the mixture in a 250-ml platinum crucible, soaking it for 12 hr at 1470°C and cooling it at 1.3°C/hr to 1110°C. A polished crystal is shown in Fig. 10.9.

Growth of $Ba(Y_{0.5}Ta_{0.5})O_3$ crystals from a flux in a temperature gradient using a seed crystal suspended below the melt surface also was attempted. In this method, an excess of nutrient material (i.e. material of the composition to be grown) is available in the bottom of the crucible, which is maintained at a higher temperature than the surface of the melt. When the seed crystal is placed in the cooler region of the crucible, some of the nutrient dissolves, is transported by diffusion through the flux, and deposits on the rotating seed as it is slowly withdrawn from the melt.

In these experiments a 400-g charge of composition $BaO:YTaO_4:B_2O_3 = 52:35:13$ was equilibrated for 20 hr. The melt surface was held 8°C cooler than the bottom of the crucible. The seed was lowered into the melt, rotated at about 100 rev/min and withdrawn at a rate of 0.0025 in./hr. In a 4-day growth period, the linear growth rate was of the order of 0.025 mm/hr. The material which was grown showed large single crystal regions; however, with better control of the volatilization of B_2O_3 large single crystals could probably be grown.

Another technique[56] considered was one developed with J. Davis, also of the United Aircraft Research Laboratories, to grow Al_2O_3 single crystals. The apparatus (see Fig. 10.10) is similar to an electron beam zone melter except that it requires no anode wire around the insulator and it can be used in partial pressures of oxygen or other gases. The material to be used is made into a polycrystalline rod and passed through a ring (cold cathode) which impinges electrons on a small zone and melts it. As the molten zone passes along the material it purifies the rod as it transforms it into a single crystal. The technique seems to be well suited for high melting point oxides.

$La(B^{2+}_{0.5}B^{4+}_{0.5})O_3$

Single crystals of $La(B^{2+}_{0.5}B^{4+}_{0.5})O_3$-type compounds where B^{2+} is Ni, Mg, Zn and B^{4+} is Ir or Ru were grown from fluxes in platinum crucibles[55]. The amount and composition of the fluxes, firing conditions and cooling rates are listed in Table 10.3.

The crystals produced in this manner were found to be electrically conducting.

Fig. 10.9. Ba(Y$_{0.5}$Ta$_{0.5}$)O$_3$ single crystals grown from a flux (after Galasso et al.[56]).

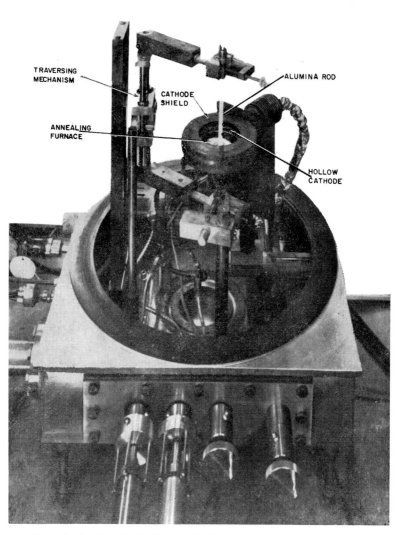

Fig. 10.10. Annular hollow cathode crystal-growing apparatus.

TABLE 10.3. *Crystal Growth Data for* $La(B_{0.5}^{2+}B_{0.5}^{4+})O_3$-*type Compounds (after Galasso and Darby*[55])

Compound	Flux	Flux (wt%)	Temp. range (°C)	Cooling rate (°C/hr)	Crystal size (mm) on edge
$La(Mg_{0.5}Ru_{0.5})O_3$	PbO–PbF$_2$	85	1320–1000	30	1.0
$La(Ni_{0.5}Ir_{0.5})O_3$	PbO–PbF$_2$	85	1300–1000	30	0.5
$La(Ni_{0.5}Ru_{0.5})O_3$	PbO–PbF$_2$	85	1300–1000	30	2.0
	PbO	85	1300–850	30	0.5
$La(Zn_{0.5}Ru_{0.5})O_3$	PbO	85	1300–25	30	0.1

REFERENCES

1. E. J. FRESIA, L. KATZ and R. WARD, *J. Am. Chem. Soc.* **81**, 4783 (1959).
2. F. GALASSO, L. KATZ and R. WARD, *J. Am. Chem. Soc.* **81**, 820 (1959).
3. A. SLEIGHT and R. WARD, *Inorg. Chem.* **1**, 790 (1962).
4. A. SLEIGHT, J. LONGO and R. WARD, *Inorg. Chem.* **1**, 245 (1962).
5. F. PATTERSON, C. W. MOELLER and R. WARD, *Inorg. Chem.* **2**, 196 (1963).
6. D. RIDGLEY and R. WARD, *J. Am. Chem. Soc.* **77**, 6135 (1955).
7. W. MCCARROLL, R. WARD and L. KATZ, *J. Am. Chem. Soc.* **78**, 2909 (1956).
8. J. RANDALL and R. WARD, *J. Am. Chem. Soc.* **81**, 2629 (1959).
9. M. KESTIGIAN, J. G. DICKENSON and R. WARD, *J. Am. Chem. Soc.* **79**, 5598 (1957).
10. F. SUGAWARA and S. IIDA, *J. Phys. Soc. Japan* **20**, 1529 (1965).
11. C. FELDMAN, *Rev. Sci. Instr.* **26**, 463 (1955).
12. R. C. DEVRIES, *J. Am. Ceram. Soc.* **45**, 225 (1962).
13. J. T. LAST, *Rev. Sci. Instr.* **28**, 720 (1959).
14. H. PFISTEIER, E. FUCHS and W. LIESK, *Naturwiss.* **49**, 178 (1962).
15. F. V. BURSIAN and N. P. SMIRNOVA, *Fiz. Tverd. Tela.* **4**, 1675 (1962).
16. C. FELDMAN, *Rev. Sci. Instr.* **26**, 463 (1955).
17. J. P. GREEN, *A Method for Fabricating Thin Ferroelectric Films of* $BaTiO_3$, Tech. Mem. ESL-TM-105, MIT (April 1961).

18. D. FRANKL, A. HAGENLOCKER, E. O. HAFFNER, P. H. KLECK, A. SANDOR, E. BOTH and H. J. DEGENHART, *Proc. Electron Components, Conf.* **44**, (1962).
19. A. MOLL, Z. *Angew. Phys.* **10**, 410 (1958).
20. O. RODER, Z. *Angew. Phys.* **12**, 323 (1960).
21. E. K. MÜLLER, B. J. NICHOLSON and M. H. FRANCOMBE, *Electrochem. Techn.* **1**, 158 (1963).
22. E. K. MÜLLER, B. J. NICHOLSON and G. TURNER, *J. Electrochem. Soc.* **110**, 969 (1963).
23. *Electronics*, P. 54 (1 Sept. 1961).
24. B. T. MATTHIAS and J. P. REMEIKA, *Phys. Rev.* **76**, 1886 (1949).
25. G. SHIRANE, H. DANNER, A. PAVLOVIC and R. PEPINSKY, *Phys. Rev.* **93**, 612 (1954).
26. S. TRIEBWASSER, *Phys. Rev.* **101**, 993 (1956).
27. C. PULVARI, WADD Technical Report 60–146, Cath. Univ. of Am. (April 1960).
28. L. E. CROSS and B. J. NICHOLSON, *Phil. Mag.* **46**, 453 (1955).
29. H. F. KAY, *Report Prog. Phys.* **18**, 230 (1955).
30. S. H. WEMPLE, Ph.D. Thesis, MIT (1963).
31. W. A. BONNER, E. F. DEARBORN and L. G. VAN UITERT, *Cer. Bulletin* **44**, 19 (1965).
32. H. BLATTNER, B. MATTHIAS, and W. MERZ, *Helv. Chim. Acta* **20**, 225 (1947).
33. J. P. REMEIKA, *J. Am. Chem. Soc.* **76**, 940 (1954).
34. R. C. LINARES, *J. Am. Chem. Soc.* **64**, 941 (1942).
35. R. C. DEVRIES, *J. Am. Ceram. Soc.* **42**, 547 (1959).
36. A. VON HIPPEL et al., Technical Report 178, Lab. Ins. Res., MIT (March 1963).
37. F. BROWN and W. H. TOAT, *J. Appl. Phys.* **35**, 1594 (1964).
38. R. DEVRIES, *J. Am. Ceram. Soc.* **47**, 134 (1964).
39. H. F. KAY, Report L/S 257, Brit. Elec. Res. Assoc. (1951).
40. L. MERKER, *J. Am. Ceram. Soc.* **45**, 366 (1962).
41. L. MERKER, *Mining Eng.* **202**, 647 (1955).
42. H. H. ROGERS, Tech. Report No. 56, Lab. for Inst. Res. MIT (1952).
43. S. NOMURA and S. SAWADA, *Report Inst. Sci. Tech., Univ. Tokyo* **6**, 11 (1952).
44. H. F. KAY and J. L. MILES, Report L/T 303, Brit. Elec. Res. Assoc., 5 (1954).
45. C. PULVARI, WADC Technical Note 56–467, Cath. Univ. of Am. (Feb. 1957).
46. F. JONA, G. SHIRANE and R. PEPINSKY, *Phys. Rev.* **97**, 1585 (1955).
47. J. P. REMEIKA, *J. Am. Chem. Soc.* **78**, 4258 (1956).
48. Airtron Semiannual Tech. Summ. Rpt. NONR–4616(00) (1 July 1965–31 Dec. 1965).
49. C. D. BRANDLE, H. FAY and O. H. NESTOR, Final Rpt. (Jan. 1965–May 1965) NONR–4793(00).
50. R. C. OHLMANN, R. MAZELSKY and J. MURPHY, Final Tech. Rpt. (16 Apr. 1964–16 Oct. 1965), NONR–4658(00).

51. M. E. STRAUMANIS, *J. Am. Chem. Soc.* **71,** 679 (1949).
52. V. A. BOKOV and I. E. MYL'NIKOVA, *Sov. Phys., Solid State* **2,** 2428 (1961).
53. V. A. BOKOV, S. A. KIZHAEV, I. E. MYL'NIKOVA and A. G. TUTOV, *Sov. Phys., Solid State* **6,** 2419 (1965).
54. F. GALASSO and J. PINTO, *Nature* **207,** 70 (1965).
55. F. GALASSO and W. DARBY, *Inorg. Chem.* **4,** 71 (1965).
56. F. GALASSO, G. LAYDEN and D. FINCHBAUGH, *J. Chem. Phys.* **44,** 2703 (1966).

CHAPTER 11
OTHER PEROVSKITE-TYPE COMPOUNDS

THE oxides are by far the most numerous and most interesting materials with the perovskite structure. However, there are some carbides, halides, hydrides and nitrides with this structure which have been investigated because of their magnetic properties and possible use as hosts for transition metal activating ions. Some of the data on their preparation, structure and properties is presented in this chapter.

11.1. PREPARATION OF PEROVSKITE-TYPE PHASES (OTHER THAN OXIDES)

The ternary carbides with the perovskite structure have been prepared predominantly by two techniques. The first involves melting the appropriate proportions of the two metals and carbon under argon and cooling them. Whenever pronounced coring exists, the samples are annealed for long periods.[1] The second technique is a solid-state reaction between the reactants after they are placed in an evacuated sealed silica capsule.[2] In a variation of this method, one metal is heated with carbon and the alloy is ground, mixed with the other metal and heated again.

Perovskite-type fluorides have been prepared by precipitation from aqueous solutions. However, crystals prepared in this manner are not stoichiometric.[3] Therefore, other techniques are used when possible. The halides $K(Na_{0.5}Cr_{0.5})F_3$, $K(Na_{0.5}Fe_{0.5})F_3$ and $K(Na_{0.5}Ga_{0.5})F_2$ were formed by reacting the trifluorides with KHF_3 and $NaHF_2$ in a platinum crucible over an open flame.[4] The melt was cooled, leached with water and ethyl alcohol, and the product dried.

Many fluorides with the perovskite structure also have been prepared by a solid-state reaction between an alkali metal halide and a divalent metal oxide at 500–800°C.[5] The same type of reactions have been conducted between the binary fluorides.

Single crystals of a number of fluorides have been prepared by Knox using fluorides precipitated from aqueous solutions as the reactants.[6] The fluorides were heated in anhydrous HF, mixed with KHF_2, melted and slowly cooled in an inert atmosphere. Crystals of $KMnF_3$, $KFeF_3$, $KCoF_3$, $KNiF_3$, $KCuF_3$ and $KZnF_3$ have been grown in this manner.

Single crystals of $KMnF_3$ have been grown by the Czochralski technique by Nassau.[7] In these studies, special equip-

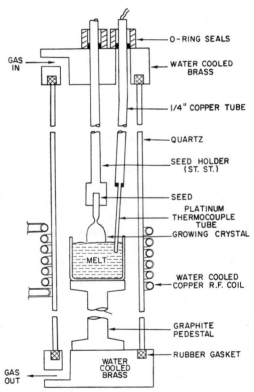

FIG. 11.1. Apparatus for Czochralski-pulling (after Nassau[7]).

ment had to be fabricated to prevent any oxygen from reacting with the melt (see Fig. 11.1).

Kestigian et al.[8] grew crystals of $RbFeF_3$ and $CsFeF_3$ using a horizontal Bridgman technique. Anhydrous fluorides were used in dense graphite containers under an HF–argon atmosphere. Single crystals 2×0.5 in. in dimensions were grown by this method.

The ternary hydrides, $LiBaH_3$ and $LiSrH_3$, were prepared by placing mixtures of the metals in a stainless-steel boat in a stainless-steel bomb and heating the mixture under a hydrogen atmosphere.[9] The metals were ground and treated in a dry box under an argon atmosphere.

The nitrides Fe_3NiN, Fe_3PtN, Fe_4N and Fe_3PdN were prepared by Stadelmaier and Fraker[10] using an induction unit to melt the alloy, which they then nitrided. The alloy was ground into powders and a mixture of NH_3 and hydrogen was used as the nitriding gas. Weiner and Berger[11] found that they could obtain nitrides by first processing the ingots into strips before nitriding. If the strips were very thin, homogeneous nitrides were obtained.

11.2. Structure

In the structure of the ternary carbides described in this chapter the Al, Sn or Ga metal atoms are in the A position, the C atom in the B position, and the transition metal atoms in the oxygen atom positions of the perovskite structure. This makes the X-ray pattern of these phases look like that produced by a face centered cubic arrangement of atoms with a superstructure. In the unit cell of Mn_3AlC structure the manganese atoms are located at the face centers, the aluminum atom is at the cube corners and the carbon atom is at the body centered position.

The nitrides Fe_3NiN and Fe_3PtN also adopt the perovskite structure. Studies by Wiener and Berger indicate that there is complete ordering in the structure of these nitrides. Because of the low scattering factor of nitrogen the powder patterns of these compounds also look as though they had a structure with a face-centered cubic lattice.

TABLE 11.1. *Unit Cell Parameters for Perovskite-type Phases (other than oxides)*

Phases	a (Å)	c (Å)		References
Carbides				
$AlFe_3C$	3.719			12
$AlMn_3C$	3.869			1
Fe_3SnC	3.85			5
$GaMn_3C$	8.376			12
Mn_3ZnC	3.92			2
Halides				
$CsCaF_3$	4.522			5
$CsCdBr_3$	10.70			13
$CsCdCl_3$	5.20			13
$CsFeF_3$	6.158	14.855	hexagonal	8
$CsGeCl_3$	5.47			13, 14
$CsHgBr_3$	5.77			13
$CsHgCl_3$	5.44	8.72	tetragonal	13
$CsMgF_3$	9.39	8.39	tetragonal	5
$CsPbBr_3$	5.874			
$CsPbCl_3$	5.590	5.630	tetragonal	
$CsZnF_3$	9.90	9.05	tetragonal	5
$KCaF_3$	8.742			5
$KCdF_3$	4.293			16
$KCoF_3$	4.071			6
$KCrF_3$	4.274	4.019	tetragonal	6
$KCuF_3$	4.140	3.926	tetragonal	6
$KFeF_3$	4.120			6
$KMgF_3$	3.973			5
$KMnF_3$	4.182			6
$KNiF_3$	4.012			6
$KZnF_3$	4.055			6
$LiBaF_3$	3.996			5
$NaZnF_3$	7.76	8.75	tetragonal	5
$RbCaF_3$	4.452			5
$RbCoF_3$	4.062			15
$RbFeF_3$	4.174			8
$RbMgF_3$	8.19		$\beta = 98°30'$ monocl.	5

TABLE 11.1 (cont.)

Phases	d (Å)	c (Å)		References
$RbMnF_3$	4.250			17
$RbZnF_3$	8.71	8.01	tetragonal	5
$K(Cr_{0.5}Na_{0.5})F_3$	8.266			4
$K(Fe_{0.5}Na_{0.5})F_3$	8.323			4
$K(Ga_{0.5}Na_{0.5})F_3$	8.246			4
Hydrides				
$LiBaH_3$	4.023			9
$LiEuH_3$	3.796			18
$LiSrH_3$	3.833			9
Nitrides				
Fe_4N	3.795			11
Mn_4N	3.857			11
Fe_3NiN	3.790			11
Fe_3PtN	3.857			11

As can be seen in Table 11.1, many of the fluorides have the "ideal" cubic perovskite structure. Of the series with the formula $KBF_3(B^{2+} = Mn$, Fe, Co, Ni, Cu, Cr and Zn), only $KCuF_3$ and $KCrF_3$ have distorted structures. Many of the others have structures related to the various modification of the perovskite type.

The complex fluorides $K(Na_{0.5}Cr_{0.5})F_3$ and $K(Na_{0.5}Fe_{0.5})F_3$ were found to adopt the ordered perovskite structure of the $(NH_4)_3FeF_6$ type. The fluoride ions in this structure move closer to the transition metal ion and away from the sodium ion.

The ternary hydrides $LiBaH_3$ and $LiSrH_3$ have the inverse perovskite structure with the lithium ions in the B position and alkaline earth metal ions in the A position. The crystallographic data for these phases as well as for the carbides, halides and nitrides are listed in Table 11.1.

11.3. Properties

Many of the carbides and nitrides with the perovskite structure are ferromagnetic materials. The carbide Mn_3AlC has been studied extensively. Butters and Myers[2] found that it was strongly magnetic at low temperature and has a Curie temperature of 15°C. They also found that Mn_3ZnC was ferromagnetic with a Curie temperature of 80°C. The Curie temperatures of both of these materials varied with the Mn/Al or Zn ratio. Figure 11.2 shows the variation of Curie

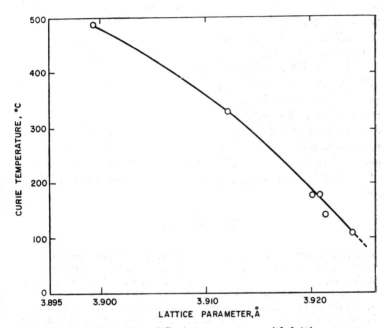

FIG. 11.2. Variation of Curie temperature with lattice parameters (after Butters and Myers[2])

temperature with lattice constant for the Mn_xZn_yC phases. Note that the Curie temperature increases as the lattice constant decreases.

The electrical resistivity of Mn_3ZnC is 770 ×cat⁶ ohm-cm at 20° and decreases with increasing temperau1.0 It is not stable in moist air and must be kept in a desictor.re

The nitrides Fe_4N, Mn_4N, Fe_3NiN and Fe_3PtN were also found to be ferromagnetic with Curie points of 488°, 465°, 487° and 369°C, respectively.

Studies by Machin et al.[19] showed that many of the ternary perovskite-type fluorides also had interesting magnetic properties. The fluorides $KFeF_3$, $KCoF_3$ and $KNiF_3$ are antiferromagnetic with Néel points of 112°, 135° and 280°K. The evidence for antiferromagnetism below 80°K for $KMnF_3$ and 215°K for $KCuF_3$ appears to be less certain. However, a study of the absorption spectra of $KMnF_3$ as a function of temperature does show anomalies in the maxima at 184°K and at 88°K. The first the authors attribute to a phase transformation and the latter to electron spin coupling.

Other fluorides which were found to be antiferromagnetic are $RbFeF_3$[20] and $RbMnF_3$,[21] with Néel temperatures of 75° and 83°K respectively. Small changes in the fluorescent properties in the vicinity of the Néel temperature have been reported for the latter compound and $KMnF_3$.[22, 23] At about half the Néel temperatures larger and more strongly temperature-dependent changes were observed in the fluorescence of these materials. This effect may involve the coupling of the lattice and magnetic interaction of the excited Mn^{2+} ion.

REFERENCES

1. R. G. BUTTERS and H. P. MYERS, *Phil. Mag.* **46**, 895 (1955).
2. R. G. BUTTERS and H. P. MYERS, *Phil. Mag.* **46**, 132 (1955).
3. W. G. PALMER, *Experimental Inorganic Chemistry*, Cambridge University Press (1954).
4. K. KNOX and D. W. MITCHELL, *J. Nucl. Chem.* **21**, 253 (1961).
5. W. L. W. LUDEKENS and A. J. E. WELCH, *Acta Cryst.* **5**, 841 (1952).
6. K. KNOX, *Acta Cryst.* **14**, 583 (1961).
7. K. NASSAU, *J. Appl. Phys.* **32**, 1820 (1961).
8. M. KESTIGIAN, F. D. LEIPZIGER, W. J. CROFT and R. GUIDOBONI, *Inorg. Chem.* **5**, 1462 (1966).
9. C. E. MESSER, J. C. EASTMAN, R. G. MERS and A. J. MAELAND, *Inorg. Chem.* **3**, 776 (1964).
10. H. H. STADELMAIER and A. C. FRAKER, *Trans. AIME* **218**, 571 (1960).
11. G. W. WIENER and J. A. BERGER, *J. Metals* **7**, 360 (1955).
12. H. P. MYERS, University of British Columbia (1956).

13. I. NÁRAY-SZABÓ, *Mulgyet Zoylemen* **1**, 30 (1947).
14. A. N. CHRISTENSEN, *Acta Chem. Scand.* **19**, 42 (1965).
15. W. RUDORFF, J. KANDLER, G. LINCKE and D. BABEL, *Angew. Chem.* **71**, 672 (1959).
16. C. BRISI, *Ann. Chem.* **42**, 356 (1952).
17. A. OKAZAKI and Y. SUEMUNE, *J. Phys. Soc. Japan* **17**, 204 (1962).
18. C. E. MESSER and K. HARDCASTLE, *Inorg. Chem.* **3**, 1327 (1964).
19. D. J. MACHIN, R. L. MARTIN and R. S. NYHOLM, *J. Chem. Soc.* **281**, 1490 (1963).
20. F. F. Y. WANG and M. KESTIGIAN, *J. Appl. Phys.* **37**, 975 (1966).
21. V. L. MORUZZI and D. T. TEANEY, *Bull. Am. Phys. Soc.* **9** (1964).
22. W. W. HOLLOWAY, JR., M. KESTIGIAN, R. NEWMAN and E. W. PROHOFSKY, *Phys. Rev. Letters* **11**, 82 (1963).
23. W. W. HOLLOWAY, JR., E. W. PROHOFSKY and M. KESTIGIAN, *Phys. Rev.* **139**, A954 (1965).

INDEX

OXIDES

$A^{1+}B^{5+}O_3$

AgNbO$_3$
  crystallography 18
  density 145
  phase transition 116,118
AgTaO$_3$
  crystallography 18
  density 145
  phase transition 116, 118
CsIO$_3$
  crystallography 5, 18
  density 145
KIO$_3$
  crystallography 5, 18
  density 145
(K,Na)NbO$_3$
  ferroelectricity 95
KNbO$_3$
  crystallography 4, 18
  density 145
  dielectric constant 87
  ferroelectricity 96
  melting point 142
  phase transition 87, 116, 117
  preparation, single crystal 165
K(Nb,Ta)O$_3$
  ferroelectricity 96
  optical properties 1
  preparation, single crystal 168
KTaO$_3$
  crystallography 5, 18
  density 145
  ferroelectricity 88
  melting point 142
  optical properties 133
  phase transition 116, 117
  preparation, single crystal 166
NaNbO$_3$
  crystallography 5, 18
  density 145
  dielectric constant 88
  ferroelectricity 87
  melting point 142
  optical properties 134
  phase transition 5, 88, 116, 117
  preparation
    single crystal 166
    thin film 164
NaTaO$_3$
  crystallography 18
  density 145
  ferroelectricity 88
  melting point 142
  phase transition 116, 118
  preparation, single crystal 166
RbIO$_3$
  crystallography 18
  density 145
RbTaO$_3$
  ferroelectricity 88
TlIO$_3$
  crystallography 5, 18
  density 145

$A^{2+}B^{4+}O_3$

(Ba,Ca,Sr)TiO$_3$
  electrical conductivity 70
(Ba,Ca,Sr)(Ti,Zr)O$_3$
  dielectric constant 92
  ferroelectricity 92
(Ba,Ca)TiO$_3$
  dielectric constant 92
  piezoelectricity 92
(Ba,Ce,Mg)TiO$_3$
  electrical conductivity 70
BaCeO$_3$
  crystallography 18
  density 145
  preparation, thin film 164
BaFeO$_3$
  crystallography 18
  density 145
Ba(Hf,Ti)O$_3$
  dielectric constant 94
  ferroelectricity 94
  phase transition 94

(Ba,La)(Mn,Ti)O_3
  ferromagnetism 123
  phase transition 123
(Ba,La)TiO_3
  electrical conductivity 70
  thermoelectricity 76
BaMoO_3
  crystallography 18
  density 145
BaPbO_3
  crystallography 18
  density 145
(Ba,Pb)(Sn,Ti)O_3
  ferroelectricity 95
(Ba,Pb)TiO_3
  dielectric constant 91
  electrical conductivity 70
  ferroelectricity 91
  piezoelectricity 98
(Ba,Pb)ZrO_3
  dielectric constant 98
  ferroelectricity 98
BaPrO_3
  crystallography 18
  density 145
BaPuO_3
  crystallography 18
  density 145
Ba(Si,Ti)O_3
  electrical conductivity 70
BaSnO_3
  crystallography 7, 18
  density 145
  preparation, thin film 164
Ba(Sn,Ti)O_3
  ferroelectricity 94
(Ba,Sr)(Sn,Ti)O_3
  electrical conductivity 70
(Ba,Sr)TiO_3
  crystallography 90
  dielectric constant 90
  electrical conductivity 90
  ferroelectricity 96
  superconductivity 64
BaThO_3
  crystallography 18
  density 145
  melting point 142
BaTiO_3
  crystallography 6, 18
  density 145
  dielectric constant 82
  electrical conductivity 68
  electron paramagnetic resonance 57
  ferroelectricity 80
  heat of formation 142
  melting point 142
  nuclear irradiation 104
  optical properties 129
  phase transition 6, 80, 115
  piezoelectricity 110
  preparation
    single crystal 169
    thin film 162
  thermal conductivity 142
  X-ray diffraction 50
Ba(Ti,Zr)O_3
  electrical conductivity 70
BaUO_3
  crystallography 18
  density 145
BaZrO_3
  crystallography 6, 18
  density 145
  melting point 142
  thermal conductivity 142
CaCeO_3
  crystallography 18
  density 145
CaHfO_3
  crystallography 18
  density 145
  melting point 142
  thermal expansion 142
(Ca,La)MnO_3
  ferromagnetism 122
  thermoelectricity 76
CaMnO_3
  crystallography 18
  density 145
  electrical conductivity 68
CaMoO_3
  crystallography 18
  density 145
  electrical conductivity 64
  preparation, powder 161
CaSnO_3
  crystallography 19
  density 145
(Ca,Sr)TiO_3
  superconductivity 64
CaThO_3
  crystallography 19
  density 145
CaTiO_3
  crystallography 3, 5, 19
  density 145
  dielectric constant 87
  electrical conductivity 68

INDEX 193

CaTiO$_3$ (cont.)
  heat of formation 142
  Madelung constant 40
  melting point 142
  optical properties 132
  phase transition 87, 117
  preparation
    single crystal 173
    thin film 164
  thermal conductivity 142
  thermal expansion 142
CaUO$_3$
  crystallography 7, 19
  density 145
CaVO$_3$
  crystallography 19
  density 145
CaZrO$_3$
  crystallography 7, 19
  density 145
  melting point 142
CdCeO$_3$
  crystallography 19
  density 145
CdSnO$_3$
  crystallography 19
  density 145
CdThO$_3$
  crystallography 19
  density 145
CdTiO$_3$
  crystallography 19
  density 145
  ferroelectricity 104
  preparation, single crystal 174
CdZrO$_3$
  crystallography 19
  density 145
EuTiO$_3$
  crystallography 19
  density 145
MgTiO$_3$
  crystallography 19
  density 145
PbCeO$_3$
  crystallography 19
  density 146
PbHfO$_3$
  crystallography 19
  density 146
  ferroelectricity 89
  phase transition 89, 116, 117
PbSnO$_3$
  crystallography 19
  density 146

PbTiO$_3$
  crystallography 6, 19
  density 146
  dielectric constant 85
  ferroelectricity 85
  phase transition 6, 85, 116, 117
  piezoelectricity 85
  preparation
    single crystal 174
    thin film 165
  thermal expansion 142
  X-ray diffraction 50
PbTiO$_3$–KNbO$_3$
  dielectric constant 95
PbTiO$_3$–NaNbO$_3$
  ferroelectricity 96
Pb(Ti,Zr)O$_3$
  dielectric constant 97
  ferroelectricity 97
  mechanical properties 158
  piezoelectricity 97, 110
PbZrO$_3$
  crystallography 7, 19
  density 146
  dielectric constant 88
  ferroelectricity 88
  phase transition 88, 116, 117
  preparation, single crystal 174
SrCeO$_3$
  crystallography 19
  density 146
SrCoO$_3$
  crystallography 19
  density 146
SrFeO$_3$
  crystallography 19
  density 146
SrHfO$_3$
  crystallography 19
  density 146
SrMoO$_3$
  crystallography 19
  density 146
  electrical conductivity 64
SrPbO$_3$
  crystallography 19
  density 146
SrRuO$_3$
  crystallography 19
  density 146
  preparation, powder 161
SrSnO$_3$
  crystallography 19
  density 146

194   INDEX

$SrSnO_3$ *(cont.)*
  preparation, thin film 164
$SrThO_3$
  crystallography 19
  density 146
$SrTiO_3$
  crystallography 6, 19
  density 146
  dielectric constant 86
  electrical conductivity 68
  electron paramagnetic resonance 58
  ferroelectricity 86
  heat of formation 142
  melting point 142
  optical properties 130
  phase transition 116, 117
  preparation, single crystal 173
  thermal conductivity 142
  thermal expansion 142
$SrUO_3$
  crystallography 19
  density 146
$SrZrO_3$
  crystallography 7, 19
  density 146
  mechanical properties 144
  melting point 142
  thermal expansion 142

$A^{3+}B^{3+}O_3$

$BiAlO_3$
  crystallography 19
  density 146
$BiCrO_3$
  crystallography 20
  density 146
  preparation, powder 162
$BiMnO_3$
  crystallography 20
  density 146
  ferromagnetism 126
  preparation, powder 162
$CeAlO_3$
  crystallography 20
  density 146
$CeCrO_3$
  crystallography 20
  density 146
$CeFeO_3$
  crystallography 20
  density 146

$CeGaO_3$
  crystallography 20
  density 146
$CeScO_3$
  crystallography 20
  density 146
$CeVO_3$
  crystallography 20
  density 146
$CrBiO_3$
  crystallography 20
  density 146
$DyAlO_3$
  crystallography 20
  density 146
$DyFeO_3$
  crystallography 20
  density 146
$DyMnO_3$
  crystallography 20
  density 146
$EuAlO_3$
  crystallography 9, 20
  density 146
$EuCrO_3$
  crystallography 20
  density 146
$EuFeO_3$
  crystallography 9, 20
  density 146
$FeBiO_3$
  crystallography 20
  density 146
$GdAlO_3$
  crystallography 9, 20
  density 146
  melting point 142
  preparation, single crystal 175
$GdCoO_3$
  crystallography 20
  density 146
$GdCrO_3$
  crystallography 9, 20
  density 146
$GdFeO_3$
  crystallography 8, 20
  density 146
  ferromagnetism 128
  X-ray diffraction 51
$GdMnO_3$
  crystallography 20
  density 146
$GdScO_3$
  crystallography 9, 20
  density 147

INDEX

GdVO$_3$
  crystallography, 9, 20
  density 147
LaAlO$_3$
  crystallography 9, 10, 20
  density 147
  laser properties 1
  melting point 142
  phase transition 10
  preparation, single crystal 174
LaCoO$_3$
  crystallography 21
  density 147
  electrical conductivity 69
  preparation, single crystal 174
La(Co,Mn)O$_3$
  ferromagnetism 123
LaCrO$_3$
  crystallography 9, 21
  density, 147
  electrical conductivity 69
La(Cr,Mn)O$_3$
  ferromagnetism 122
LaFeO$_3$
  crystallography 21
  density 147
  electrical conductivity 68
  melting point 142
LaGaO$_3$
  crystallography 9, 21
  density 147
  phase transition 9
LaInO$_3$
  crystallography 21
  density 147
LaMnO$_3$
  electrical conductivity 68
La(Mn,Ni)O$_3$
  ferromagnetism 124
LaNiO$_3$
  crystallography 21
  density 147
LaRhO$_3$
  crystallography 21
  density 147
LaScO$_3$
  crystallography 9, 21
  density 147
(La,Sr)CoO$_3$
  electrical conductivity 69
  ferromagnetism 122
(La,Sr)CrO$_3$
  electrical conductivity 69

(La,Sr)FeO$_3$
  electrical conductivity 69
  thermoelectricity 76
(La,Sr)MnO$_3$
  electrical conductivity 69
LaTiO$_3$
  crystallography 21
  density 147
  electrical conductivity 60
LaVO$_3$
  crystallography 21
  density 147
  electrical conductivity 60
  preparation, powder 161
LaYO$_3$
  crystallography 21
  density 147
NdAlO$_3$
  crystallography 9, 21
  density 147
NdCoO$_3$
  crystallography 21
  density 147
NdCrO$_3$
  crystallography 9, 21
  density 147
NdFeO$_3$
  crystallography 9, 21
  density 147
NdGaO$_3$
  crystallography 9, 21
  density 147
NdInO$_3$
  crystallography 21
  density 147
NdMnO$_3$
  crystallography 21
  density 147
NdScO$_3$
  crystallography 9, 21
  density 147
NdVO$_3$
  crystallography 9, 21
  density 147
PrAlO$_3$
  crystallography 9, 21
  density 147
PrCoO$_3$
  crystallography 21
  density 147
PrCrO$_3$
  crystallography 9, 21
  density 147

PrFeO$_3$
  crystallography 9, 22
  density 147
PrGaO$_3$
  crystallography 9, 22
  density 147
PrMnO$_3$
  crystallography 22
  density 147
PrScO$_3$
  crystallography 9, 22
  density 147
PrVO$_3$
  crystallography 9, 22
  density 147
PuAlO$_3$
  crystallography 22
  density 147
PuCrO$_3$
  crystallography 22
  density 147
PuMnO$_3$
  crystallography 22
  density 147
PuVO$_3$
  crystallography 22
  density 147
SmAlO$_3$
  crystallography 9, 22
  density 147
  phase transition 9
SmCoO$_3$
  crystallography 22
  density 147
SmCrO$_3$
  crystallography 9, 22
  density 147
SmFeO$_3$
  crystallography 9, 22
  density 147
SmInO$_3$
  crystallography 22
  density 147
SmVO$_3$
  crystallography 22
  density 147
YAlO$_3$
  crystallography 9, 22
  density 147
  melting point 142
YCrO$_3$
  crystallography 9, 22
  density 148

YFeO$_3$
  crystallography 9, 22
  density 148
YScO$_3$
  crystallography 9, 22
  density 148

A$_x$BO$_3$ and ABO$_{3-x}$

Ce$_{0.33}$NbO$_3$
  crystallography 22
  density 148
Ce$_{0.33}$TaO$_3$
  crystallography 22
  density 148
Dy$_{0.33}$TaO$_3$
  crystallography 23
  density 148
Gd$_{0.33}$TaO$_3$
  crystallography 23
  density 148
La$_{0.33}$NbO$_3$
  crystallography 23
  density 148
La$_{0.33}$TaO$_3$
  crystallography 23
  density 148
Nd$_{0.33}$NbO$_3$
  crystallography 23
  density 148
Nd$_{0.33}$TaO$_3$
  crystallography 23
  density 148
Pr$_{0.33}$NbO$_3$
  crystallography 23
  density 148
Pr$_{0.33}$TaO$_3$
  crystallography 23
  density 148
Sm$_{0.33}$TaO$_3$
  crystallography 23
  density 148
Y$_{0.33}$TaO$_3$
  crystallography 23
  density 148
Yb$_{0.33}$TaO$_3$
  crystallography 23
  density 148
Ca$_{0.5}$TaO$_3$
  crystallography 23
  density 148
Li$_x$WO$_3$
  catalyst 141
  crystallography 10, 23

INDEX

Li_xWO_3 *(cont.)*
  density 148
  electrical conductivity 62
Na_xWO_3
  catalyst 141
  crystallography 10, 23
  density 148
  electrical conductivity 60
  preparation, single crystal 175
Sr_xNbO_3
  crystallography 10, 23
  density 148
  preparation, powder 161
$BaTiO_{3-x}$
  electrical conductivity 66
$CaMnO_{3-x}$
  crystallography 11, 23
  density 148
$CaTiO_{3-x}$
  electrical conductivity 66
$SrCoO_{3-x}$
  crystallography 11, 23
  density 148
$SrFeO_{3-x}$
  crystallography 11, 23
  density 148
$SrTiO_{3-x}$
  crystallography 11, 23
  density 148
  electrical conductivity 66
  superconductivity 63
$SrVO_{3-x}$
  crystallography 11, 23
  density 148
  electrical conductivity 60

$A(B'_{0.67}B''_{0.33})O_3$
$Ba(Al_{0.67}W_{0.33})O_3$
  crystallography 23
  density 148
$Ba(Bi_{0.67}W_{0.33})O_3$
  dielectric constant 99
$Ba(Dy_{0.67}W_{0.33})O_3$
  crystallography 23
  density 148
$Ba(Er_{0.67}W_{0.33})O_3$
  crystallography 24
  density 148
$Ba(Eu_{0.67}W_{0.33})O_3$
  crystallography 24
  density 148
$Ba(Fe_{0.67}U_{0.33})O_3$
  crystallography 24
  density 148
$Ba(Gd_{0.67}W_{0.33})O_3$
  crystallography 24
  density 148
$Ba(In_{0.67}U_{0.33})O_3$
  crystallography 24
  density 148
$Ba(In_{0.67}W_{0.33})O_3$
  crystallography 24
  density 148
$Ba(La_{0.67}W_{0.33})O_3$
  crystallography 24
  density 148
$Ba(Lu_{0.67}W_{0.33})O_3$
  crystallography 24
  density 148
$Ba(Nd_{0.67}W_{0.33})O_3$
  crystallography 24
  density 148
$Ba(Sc_{0.67}U_{0.33})O_3$
  crystallography 24
  density 148
$Ba(Sc_{0.67}W_{0.33})O_3$
  crystallography 11, 24
  density 149
$Ba(Y_{0.67}U_{0.33})O_3$
  crystallography 24
  density 149
$Ba(Y_{0.67}W_{0.33})O_3$
  crystallography 24
  density 149
$Ba(Yb_{0.67}W_{0.33})O_3$
  crystallography 24
  density 149
$La(Co_{0.67}Nb_{0.33})O_3$
  crystallography 24
  density 149
$La(Co_{0.67}Sb_{0.33})O_3$
  crystallography 24
  density 149
$Pb(Fe_{0.67}W_{0.33})O_3$
  crystallography 24
  density 148
  ferroelectricity 100
$Sr(Cr_{0.67}Re_{0.33})O_3$
  crystallography 13, 24
  density 149
$Sr(Cr_{0.67}U_{0.33})O_3$
  crystallography 24
  density 149
$Sr(Fe_{0.67}Re_{0.33})O_3$
  crystallography 24
  density 149
$Sr(Fe_{0.67}W_{0.33})O_3$
  crystallography 24
  density 149

INDEX

$Sr(In_{0.67}Re_{0.33})O_3$
crystallography 24
density 149

$A^{2+}(B^{2+}_{0.33}B^{5+}_{0.67})O_3$

$Ba(Ca_{0.33}Nb_{0.67})O_3$
crystallography 25
density 149
$Ba(Ca_{0.33}Ta_{0.67})O_3$
crystallography 25
density 149
preparation, single crystal 177
$Ba(Cd_{0.33}Nb_{0.67})O_3$
crystallography 25
density 149
$Ba(Cd_{0.33}Ta_{0.67})O_3$
crystallography 25
density 149
$Ba(Co_{0.33}Nb_{0.67})O_3$
crystallography 25
density 149
$Ba(Co_{0.33}Ta_{0.67})O_3$
crystallography 25
density 149
$Ba(Cu_{0.33}Nb_{0.67})O_3$
crystallography 25
density 149
$Ba(Fe_{0.33}Nb_{0.67})O_3$
crystallography 25
density 149
$Ba(Fe_{0.33}Ta_{0.67})O_3$
crystallography 25
density 149
preparation, powder 160
$Ba(Mg_{0.33}Nb_{0.67})O_3$
crystallography 25
density 149
$Ba(Mg_{0.33}Ta_{0.67})O_3$
crystallography 25
density 149
preparation, single crystal 177
X-ray diffraction 55
$Ba(Mn_{0.33}Nb_{0.67})O_3$
crystallography 25
density 149
$Ba(Mn_{0.33}Ta_{0.67})O_3$
crystallography 25
density 149
$Ba(Ni_{0.33}Nb_{0.67})O_3$
crystallography 25
density 149

$Ba(Ni_{0.33}Ta_{0.67})O_3$
crystallography 25
density 150
preparation, single crystal 177
$Ba(Pb_{0.33}Nb_{0.67})O_3$
crystallography 25
density 150
$Ba(Pb_{0.33}Ta_{0.67})O_3$
crystallography 25
density 150
$Ba(Sr_{0.33}Ta_{0.67})O_3$
crystallography 13, 15, 25
density 150
X-ray diffraction 56
$Ba(Zn_{0.33}Nb_{0.67})O_3$
crystallography 15, 25
density 150
$Ba(Zn_{0.33}Ta_{0.67})O_3$
crystallography 25
density 150
preparation, single crystal 177
$Ca(Ni_{0.33}Nb_{0.67})O_3$
crystallography 25
density 150
$Ca(Ni_{0.33}Ta_{0.67})O_3$
crystallography 25
density 150
$Pb(Co_{0.33}Nb_{0.67})O_3$
crystallography 25
density 150
dielectric constant 102
ferroelectricity 102
preparation, single crystal 176
$Pb(Co_{0.33}Ta_{0.67})O_3$
crystallography 25
density 150
dielectric constant 102
ferroelectricity 102
preparation, single crystal 176
$Pb(Mg_{0.33}Nb_{0.67})O_3$
crystallography 26
density 150
dielectric constant 101
ferroelectricity 101
$Pb(Mg_{0.33}Ta_{0.67})O_3$
crystallography 26
density 150
dielectric constant 102
ferroelectricity 102
preparation, single crystal 177
$Pb(Mn_{0.33}Nb_{0.67})O_3$
crystallography 26
density 150

$Pb(Ni_{0.33}Nb_{0.67})O_3$
  crystallography 26
  density 150
  dielectric constant 101
  ferroelectricity 101
$Pb(Ni_{0.33}Ta_{0.67})O_3$
  crystallography 26
  density 150
  dielectric constant 102
  ferroelectricity 102
  preparation, single crystal 176
$Pb(Zn_{0.33}Nb_{0.67})O_3$
  crystallography 26
  density 150
  dielectric constant 102
  ferroelectricity 102
  preparation, single crystal 176
$Sr(Ca_{0.33}Nb_{0.67})O_3$
  crystallography 26
  density 150
$Sr(Ca_{0.33}Sb_{0.67})O_3$
  crystallography 26
  density 150
$Sr(Ca_{0.33}Ta_{0.67})O_3$
  crystallography 26
  density 150
$Sr(Cd_{0.33}Nb_{0.67})O_3$
  crystallography 26
  density 150
$Sr(Co_{0.33}Nb_{0.67})O_3$
  crystallography 26
  density 150
$Sr(Co_{0.33}Sb_{0.67})O_3$
  crystallography 26
  density 150
$Sr(Co_{0.33}Ta_{0.67})O_3$
  crystallography 26
  density 150
$Sr(Cu_{0.33}Sb_{0.67})O_3$
  crystallography 26
  density 150
$Sr(Fe_{0.33}Nb_{0.67})O_3$
  crystallography 26
  density 150
$Sr(Mg_{0.33}Nb_{0.67})O_3$
  crystallography 26
  density 150
$Sr(Mg_{0.33}Sb_{0.67})O_3$
  crystallography 26
  density 150
$Sr(Mg_{0.33}Ta_{0.67})O_3$
  crystallography 26
  density 150

$Sr(Mn_{0.33}Nb_{0.67})O_3$
  crystallography 26
  density 150
$Sr(Mn_{0.33}Ta_{0.67})O_3$
  crystallography 26
  density 150
$Sr(Ni_{0.33}Nb_{0.67})O_3$
  crystallography 26
  density 150
$Sr(Ni_{0.33}Ta_{0.67})O_3$
  crystallography 26
  density 150
$Sr(Pb_{0.33}Nb_{0.67})O_3$
  crystallography 26
  density 150
$Sr(Pb_{0.33}Ta_{0.67})O_3$
  crystallography 26
  density 150
$Sr(Zn_{0.33}Nb_{0.67})O_3$
  crystallography 27
  density 150
$Sr(Zn_{0.33}Ta_{0.67})O_3$
  crystallography 27
  density 150

$A^{2+}\left(B^{3+}_{0.5}B^{5+}_{0.5}\right)O_3$

$Ba(Bi_{0.5}Mo_{0.5})O_3$
  dielectric constant 99
$Ba(Bi_{0.5}Nb_{0.5})O_3$
  crystallography 27
  density 150
  dielectric constant 99
  ferroelectricity 99
$Ba(Bi_{0.5}Ta_{0.5})O_3$
  crystallography 27
  density 150
  dielectric constant 99
  ferroelectricity 99
$Ba(Bi_{0.5}U_{0.5})O_3$
  dielectric constant 99
$Ba(Ce_{0.5}Nb_{0.5})O_3$
  crystallography 27
  density 150
$Ba(Ce_{0.5}Pa_{0.5})O_3$
  crystallography 27
  density 150
$Ba(Co_{0.5}Nb_{0.5})O_3$
  crystallography 27
  density 150

$Ba(Co_{0.5}Re_{0.5})O_3$
  crystallography 27
  density 150
$Ba(Cr_{0.5}Os_{0.5})O_3$
  crystallography 27
  density 150
$Ba(Cr_{0.5}Re_{0.5})O_3$
  crystallography 27
  density 150
$Ba(Cr_{0.5}U_{0.5})O_3$
  crystallography 27
  density 150
$Ba(Cu_{0.5}W_{0.5})O_3$
  crystallography 27
  density 150
$Ba(Dy_{0.5}Nb_{0.5})O_3$
  crystallography 27
  density 150
$Ba(Dy_{0.5}Pa_{0.5})O_3$
  crystallography 27
  density 150
$Ba(Dy_{0.5}Ta_{0.5})O_3$
  crystallography 27
  density 150
$Ba(Er_{0.5}Nb_{0.5})O_3$
  crystallography 27
  density 150
$Ba(Er_{0.5}Pa_{0.5})O_3$
  crystallography 27
  density 150
$Ba(Er_{0.5}Re_{0.5})O_3$
  crystallography 27
  density 150
$Ba(Er_{0.5}Ta_{0.5})O_3$
  crystallography 27
  density 150
$Ba(Er_{0.5}U_{0.5})O_3$
  crystallography 27
  density 150
$Ba(Eu_{0.5}Nb_{0.5})O_3$
  crystallography 27
  density 151
$Ba(Eu_{0.5}Pa_{0.5})O_3$
  crystallography 27
  density 151
$Ba(Eu_{0.5}Ta_{0.5})O_3$
  crystallography 28
  density 151
$Ba(Fe_{0.5}Mo_{0.5})O_3$
  crystallography 28
  density 151
  ferromagnetism 126
$Ba(Fe_{0.5}Nb_{0.5})O_3$
  crystallography 28
  density 151

$Ba(Fe_{0.5}Re_{0.5})O_3$
  crystallography 28
  density 151
  ferromagnetism 125
$Ba(Fe_{0.5}Ta_{0.5})O_3$
  crystallography
    15, 28
  density 151
  thermal expansion
    142
$Ba(Gd_{0.5}Nb_{0.5})O_3$
  crystallography 28
  density 151
  optical properties 138
$Ba(Gd_{0.5}Pa_{0.5})O_3$
  crystallography 28
  density 151
$Ba(Gd_{0.5}Re_{0.5})O_3$
  crystallography 28
  density 151
$Ba(Gd_{0.5}Sb_{0.5})O_3$
  crystallography 28
  density 151
$Ba(Gd_{0.5}Ta_{0.5})O_3$
  crystallography 28
  density 151
  laser properties 138
  preparation, single
    crystal 177
$Ba(Ho_{0.5}Nb_{0.5})O_3$
  crystallography 28
  density 151
$Ba(Ho_{0.5}Pa_{0.5})O_3$
  crystallography 28
  density 151
$Ba(Ho_{0.5}Ta_{0.5})O_3$
  crystallography 28
  density 151
$Ba(In_{0.5}Nb_{0.5})O_3$
  crystallography 28
  density 151
$Ba(In_{0.5}Os_{0.5})O_3$
  crystallography 28
  density 151
$Ba(In_{0.5}Pa_{0.5})O_3$
  crystallography 28
  density 151
$Ba(In_{0.5}Re_{0.5})O_3$
  crystallography 28
  density 151
$Ba(In_{0.5}Sb_{0.5})O_3$
  crystallography 28
  density 151
$Ba(In_{0.5}Ta_{0.5})O_3$
  crystallography 28

  density 151
  laser properties 139
$Ba(In_{0.5}U_{0.5})O_3$
  crystallography 28
  density 151
$Ba(La_{0.5}Nb_{0.5})O_3$
  crystallography 28
  density 151
$Ba(La_{0.5}Pa_{0.5})O_3$
  crystallography 28
  density 151
$Ba(La_{0.5}Re_{0.5})O_3$
  crystallography 29
  density 151
$Ba(La_{0.5}Ta_{0.5})O_3$
  crystallography
    15, 29
  density 151
  laser properties 139
  preparation, single
    crystal 177
$Ba(Lu_{0.5}Nb_{0.5})O_3$
  crystallography 29
  density 151
$Ba(Lu_{0.5}Pa_{0.5})O_3$
  crystallography 29
  density 151
$Ba(Lu_{0.5}Ta_{0.5})O_3$
  crystallography 29
  density 151
  laser properties 139
$Ba(Mn_{0.5}Nb_{0.5})O_3$
  crystallography 29
  density 151
$Ba(Mn_{0.5}Re_{0.5})O_3$
  crystallography 29
  density 151
$Ba(Mn_{0.5}Ta_{0.5})O_3$
  crystallography 29
  density 151
$Ba(Nd_{0.5}Nb_{0.5})O_3$
  crystallography 29
  density 151
$Ba(Nd_{0.5}Pa_{0.5})O_3$
  crystallography 29
  density 151
$Ba(Nd_{0.5}Re_{0.5})O_3$
  crystallography 29
  density 151
$Ba(Nd_{0.5}Ta_{0.5})O_3$
  crystallography 29
  density 151
$Ba(Ni_{0.5}Nb_{0.5})O_3$
  crystallography 29
  density 151

INDEX

$Ba(Pr_{0.5}Nb_{0.5})O_3$
  crystallography 29
  density 151
$Ba(Pr_{0.5}Pa_{0.5})O_3$
  crystallography 29
  density 151
$Ba(Pr_{0.5}Ta_{0.5})O_3$
  crystallography 29
  density 151
$Ba(Rh_{0.5}Nb_{0.5})O_3$
  crystallography 29
  density 151
$Ba(Rh_{0.5}U_{0.5})O_3$
  crystallography 29
  density 151
$Ba(Sc_{0.5}Nb_{0.5})O_3$
  crystallography 29
  density 152
$Ba(Sc_{0.5}Os_{0.5})O_3$
  crystallography 29
  density 152
$Ba(Sc_{0.5}Pa_{0.5})O_3$
  crystallography 29
  density 152
$Ba(Sc_{0.5}Re_{0.5})O_3$
  crystallography 29
  density 152
  Madelung constant 40
$Ba(Sc_{0.5}Sb_{0.5})O_3$
  crystallography 29
  density 152
$Ba(Sc_{0.5}Ta_{0.5})O_3$
  crystallography 29
  density 152
  laser properties 139
  preparation, single
    crystal 177
$Ba(Sc_{0.5}U_{0.5})O_3$
  crystallography 30
  density 152
$Ba(Sm_{0.5}Nb_{0.5})O_3$
  crystallography 30
  density 152
$Ba(Sm_{0.5}Pa_{0.5})O_3$
  crystallography 30
  density 152
$Ba(Sm_{0.5}Ta_{0.5})O_3$
  crystallography 30
  density 152
$Ba(Tb_{0.5}Nb_{0.5})O_3$
  crystallography 30
  density 152
$Ba(Tb_{0.5}Pa_{0.5})O_3$
  crystallography 30
  density 152

$Ba(Tl_{0.5}Ta_{0.5})O_3$
  crystallography 30
  density 152
$Ba(Tm_{0.5}Nb_{0.5})O_3$
  crystallography 30
  density 152
$Ba(Tm_{0.5}Pa_{0.5})O_3$
  crystallography 30
  density 152
$Ba(Tm_{0.5}Ta_{0.5})O_3$
  crystallography 30
  density 152
$Ba(Y_{0.5}Nb_{0.5})O_3$
  crystallography 30
  density 152
$Ba(Y_{0.5}Pa_{0.5})O_3$
  crystallography 30
  density 152
$Ba(Y_{0.5}Re_{0.5})O_3$
  crystallography 30
  density 152
$Ba(Y_{0.5}Ta_{0.5})O_3$
  crystallography 30
  density 152
  laser properties 139
  preparation, single
    crystal 177
  X-ray diffraction 54
$Ba(Y_{0.5}U_{0.5})O_3$
  crystallography 30
  density 152
$Ba(Yb_{0.5}Nb_{0.5})O_3$
  crystallography 30
  density 152
$Ba(Yb_{0.5}Pa_{0.5})O_3$
  crystallography 30
  density 152
$Ba(Yb_{0.5}Ta_{0.5})O_3$
  crystallography 30
  density 152
$Ca(Al_{0.5}Nb_{0.5})O_3$
  crystallography 30
  density 152
$Ca(Al_{0.5}Ta_{0.5})O_3$
  crystallography 30
  density 152
$Ca(Co_{0.5}W_{0.5})O_3$
  crystallography 30
  density 152
$Ca(Cr_{0.5}Mo_{0.5})O_3$
  crystallography 30
  density 152
  ferromagnetism 125
$Ca(Cr_{0.5}Nb_{0.5})O_3$
  crystallography 31

  density 152
$Ca(Cr_{0.5}Os_{0.5})O_3$
  crystallography 31
  density 152
$Ca(Cr_{0.5}Re_{0.5})O_3$
  crystallography 31
  density 152
  ferromagnetism 126
$Ca(Cr_{0.5}Ta_{0.5})O_3$
  crystallography 31
  density 152
$Ca(Cr_{0.5}W_{0.5})O_3$
  crystallography 31
  density 152
  ferromagnetism 126
$Ca(Dy_{0.5}Nb_{0.5})O_3$
  crystallography 31
  density 152
$Ca(Dy_{0.5}Ta_{0.5})O_3$
  crystallography 31
  density 152
$Ca(Er_{0.5}Nb_{0.5})O_3$
  crystallography 31
  density 152
$Ca(Er_{0.5}Ta_{0.5})O_3$
  crystallography 31
  density 152
$Ca(Fe_{0.5}Mo_{0.5})O_3$
  crystallography 31
  density 152
  ferromagnetism 125
$Ca(Fe_{0.5}Nb_{0.5})O_3$
  crystallography 31
  density 152
$Ca(Fe_{0.5}Re_{0.5})O_3$
  density 152
  ferromagnetism 125
$Ca(Fe_{0.5}Sb_{0.5})O_3$
  crystallography 31
  density 152
  ferromagnetism 128
$Ca(Fe_{0.5}Ta_{0.5})O_3$
  crystallography 31
  density 152
$Ca(Gd_{0.5}Nb_{0.5})O_3$
  crystallography 31
  density 152
$Ca(Gd_{0.5}Ta_{0.5})O_3$
  crystallography 31
  density 153
$Ca(Ho_{0.5}Nb_{0.5})O_3$
  crystallography 31
  density 153

202 INDEX

$Ca(Ho_{0.5}Ta_{0.5})O_3$
 crystallography 31
 density 153
$Ca(In_{0.5}Nb_{0.5})O_3$
 crystallography 31
 density 153
$Ca(In_{0.5}Ta_{0.5})O_3$
 crystallography 31
 density 153
$Ca(La_{0.5}Nb_{0.5})O_3$
 crystallography 31
 density 153
$Ca(La_{0.5}Ta_{0.5})O_3$
 crystallography 31
 density 153
$Ca(Mn_{0.5}Ta_{0.5})O_3$
 crystallography 31
 density 153
$Ca(Nd_{0.5}Nb_{0.5})O_3$
 crystallography 31
 density 153
$Ca(Nd_{0.5}Ta_{0.5})O_3$
 crystallography 31
 density 153
$Ca(Ni_{0.5}W_{0.5})O_3$
 crystallography 32
 density 153
$Ca(Pr_{0.5}Nb_{0.5})O_3$
 crystallography 32
 density 153
$Ca(Pr_{0.5}Ta_{0.5})O_3$
 crystallography 32
 density 153
$Ca(Sc_{0.5}Re_{0.5})O_3$
 crystallography 32
 density 153
$Ca(Sm_{0.5}Nb_{0.5})O_3$
 crystallography 32
 density 153
$Ca(Sm_{0.5}Ta_{0.5})O_3$
 crystallography 32
 density 153
$Ca(Tb_{0.5}Nb_{0.5})O_3$
 crystallography 32
 density 153
$Ca(Tb_{0.5}Ta_{0.5})O_3$
 crystallography 32
 density 153
$Ca(Y_{0.5}Nb_{0.5})O_3$
 crystallography 32
 density 153
$Ca(Y_{0.5}Ta_{0.5})O_3$
 crystallography 32
 density 153

$Ca(Yb_{0.5}Nb_{0.5})O_3$
 crystallography 32
 density 153
$Ca(Yb_{0.5}Ta_{0.5})O_3$
 crystallography 32
 density 153
$Pb(Fe_{0.5}Nb_{0.5})O_3$
 crystallography 32
 density 153
 ferroelectricity 99
 preparation, single
  crystal 176
$Pb(Fe_{0.5}Ta_{0.5})O_3$
 crystallography 32
 density 153
 dielectric constant 101
 ferroelectricity 101
 preparation, single
  crystal 176
$Pb(Ho_{0.5}Nb_{0.5})O_3$
 crystallography 32
 density 153
$Pb(In_{0.5}Nb_{0.5})O_3$
 crystallography 32
 density 153
 dielectric constant 100
 ferroelectricity 100
$Pb(Lu_{0.5}Nb_{0.5})O_3$
 crystallography 32
 density 153
 dielectric constant 100
 ferroelectricity 100
 phase transition 121
$Pb(Lu_{0.5}Ta_{0.5})O_3$
 crystallography 32
 density 153
 phase transition 121
$Pb(Sc_{0.5}Nb_{0.5})O_3$
 crystallography 32
 density 153
 dielectric constant 100
 ferroelectricity 100
 preparation, single
  crystal 116
$Pb(Sc_{0.5}Ta_{0.5})O_3$
 crystallography 32
 density 153
 dielectric constant 100
 preparation, single
  crystal 176

$Pb(Yb_{0.5}Nb_{0.5})O_3$
 crystallography 32
 density 153
 ferroelectricity 99
 phase transition 121
$Pb(Yb_{0.5}Ta_{0.5})O_3$
 crystallography 32
 density 153
 dielectric constant 100
 ferroelectricity 100
$Sr(Co_{0.5}Nb_{0.5})O_3$
 crystallography 32
 density 153
$Sr(Co_{0.5}Sb_{0.5})O_3$
 crystallography 32
 density 153
$Sr(Cr_{0.5}Mo_{0.5})O_3$
 crystallography 32
 density 153
 ferromagnetism 126
$Sr(Cr_{0.5}Nb_{0.5})O_3$
 crystallography 32
 density 153
$Sr(Cr_{0.5}Os_{0.5})O_3$
 crystallography 33
 density 153
$Sr(Cr_{0.5}Re_{0.5})O_3$
 crystallography 33
 density 153
$Sr(Cr_{0.5}Sb_{0.5})O_3$
 crystallography 33
 density 153
$Sr(Cr_{0.5}Ta_{0.5})O_3$
 crystallography 33
 density 153
$Sr(Cr_{0.5}W_{0.5})O_3$
 crystallography 33
 density 153
$Sr(Dy_{0.5}Ta_{0.5})O_3$
 crystallography 33
 density 153
$Sr(Er_{0.5}Ta_{0.5})O_3$
 crystallography 33
 density 154
$Sr(Eu_{0.5}Ta_{0.5})O_3$
 crystallography 33
 density 154
$Sr(Fe_{0.5}Mo_{0.5})O_3$
 crystallography 33
 density 154
 ferromagnetism 126
$Sr(Fe_{0.5}Nb_{0.5})O_3$
 crystallography 33
 density 154

INDEX

$Sr(Fe_{0.5}Nb_{0.5})O_3$ *(cont.)*
  phase transition 119
$Sr(Fe_{0.5}Re_{0.5})O_3$
  ferromagnetism 126
$Sr(Fe_{0.5}Sb_{0.5})O_3$
  crystallography 33
  density 154
  ferromagnetism 128
$Sr(Fe_{0.5}Ta_{0.5})O_3$
  crystallography 33
  density 154
  thermal expansion 142
$Sr(Ga_{0.5}Nb_{0.5})O_3$
  crystallography 33
  density 154
$Sr(Ga_{0.5}Os_{0.5})O_3$
  crystallography 33
  density 154
$Sr(Ga_{0.5}Re_{0.5})O_3$
  crystallography 33
  density 154
$Sr(Ga_{0.5}Sb_{0.5})O_3$
  crystallography 33
  density 154
$Sr(Gd_{0.5}Ta_{0.5})O_3$
  crystallography 33
  density 154
$Sr(Ho_{0.5}Ta_{0.5})O_3$
  crystallography 33
  density 154
$Sr(In_{0.5}Nb_{0.5})O_3$
  crystallography 33
  density 154
$Sr(In_{0.5}Os_{0.5})O_3$
  crystallography 33
  density 154
$Sr(In_{0.5}Re_{0.5})O_3$
  crystallography 33
  density 154
$Sr(In_{0.5}U_{0.5})O_3$
  crystallography 33
  density 154
$Sr(La_{0.5}Ta_{0.5})O_3$
  crystallography 33
  density 154
$Sr(Lu_{0.5}Ta_{0.5})O_3$
  crystallography 33
  density 154
$Sr(Mn_{0.5}Mo_{0.5})O_3$
  crystallography 33
  density 154
$Sr(Mn_{0.5}Sb_{0.5})O_3$
  crystallography 33
  density 154

  ferromagnetism 128
$Sr(Nd_{0.5}Ta_{0.5})O_3$
  crystallography 33
  density 154
$Sr(Ni_{0.5}Sb_{0.5})O_3$
  crystallography 33
  density 154
$Sr(Rh_{0.5}Sb_{0.5})O_3$
  crystallography 34
  density 154
$Sr(Sc_{0.5}Os_{0.5})O_3$
  crystallography 34
  density 154
$Sr(Sc_{0.5}Re_{0.5})O_3$
  crystallography 34
  density 154
$Sr(Sm_{0.5}Ta_{0.5})O_3$
  crystallography 34
  density 154
$Sr(Tm_{0.5}Ta_{0.5})O_3$
  crystallography 34
  density 154
$Sr(Y_{0.5}Ta_{0.5})O_3$
  phase transition 119
$Sr(Yb_{0.5}Ta_{0.5})O_3$
  crystallography 34
  density 154

$$A^{2+}(B^{2+}_{0.5}B^{6+}_{0.5})O_3$$

$Ba(Ba_{0.5}Os_{0.5})O_3$
  crystallography 34
  density 154
  Madelung constant 41
$Ba(Ba_{0.5}Re_{0.5})O_3$
  crystallography 34
  density 154
  Madelung constant 41
$Ba(Ba_{0.5}U_{0.5})O_3$
  crystallography 34
  density 154
$Ba(Ba_{0.5}W_{0.5})O_3$
  crystallography 34
  density 154
$Ba(Ca_{0.5}Mo_{0.5})O_3$
  crystallography 16, 34
  density 154
$Ba(Ca_{0.5}Os_{0.5})O_3$
  crystallography 34
  density 154

$Ba(Ca_{0.5}Re_{0.5})O_3$
  crystallography 34
  density 154
$Ba(Ca_{0.5}Te_{0.5})O_3$
  crystallography 34
  density 154
$Ba(Ca_{0.5}U_{0.5})O_3$
  crystallography 34
  density 154
$Ba(Ca_{0.5}W_{0.5})O_3$
  crystallography 16, 34
  density 154
$Ba(Cd_{0.5}Os_{0.5})O_3$
  crystallography 34
  density 154
$Ba(Cd_{0.5}Re_{0.5})O_3$
  crystallography 34
  density 154
$Ba(Cd_{0.5}U_{0.5})O_3$
  crystallography 34
  density 155
$Ba(Co_{0.5}Mo_{0.5})O_3$
  crystallography 34
  density 155
$Ba(Co_{0.5}Re_{0.5})O_3$
  crystallography 34
  density 155
$Ba(Co_{0.5}U_{0.5})O_3$
  crystallography 34
  density 155
$Ba(Co_{0.5}W_{0.5})O_3$
  crystallography 35
  density 155
  dielectric constant 100
  ferroelectricity 100
$Ba(Cr_{0.5}U_{0.5})O_3$
  crystallography 35
  density 155
$Ba(Cu_{0.5}U_{0.5})O_3$
  crystallography 35
  density 155
$Ba(Fe_{0.5}Re_{0.5})O_3$
  crystallography 35
  density 155
$Ba(Fe_{0.5}U_{0.5})O_3$
  crystallography 35
  density 155
$Ba(Fe_{0.5}W_{0.5})O_3$
  crystallography 35
  density 155
$Ba(Mg_{0.5}Os_{0.5})O_3$
  crystallography 35
  density 155

204 INDEX

$Ba(Mg_{0.5}Re_{0.5})O_3$
  crystallography 35
  density 155
$Ba(Mg_{0.5}Te_{0.5})O_3$
  crystallography 35
  density 155
$Ba(Mg_{0.5}U_{0.5})O_3$
  crystallography 35
  density 155
$Ba(Mg_{0.5}W_{0.5})O_3$
  crystallography 15, 35
  density 155
$Ba(Mn_{0.5}Re_{0.5})O_3$
  crystallography 35
  density 155
$Ba(Mn_{0.5}U_{0.5})O_3$
  crystallography 35
  density 155
$Ba(Ni_{0.5}Mo_{0.5})O_3$
  crystallography 35
  density 155
$Ba(Ni_{0.5}Re_{0.5})O_3$
  crystallography 35
  density 155
  Madelung constant 40
$Ba(Ni_{0.5}U_{0.5})O_3$
  crystallography 35
  density 155
$Ba(Ni_{0.5}W_{0.5})O_3$
  crystallography 35
  density 155
$Ba(Pb_{0.5}Mo_{0.5})O_3$
  crystallography 35
  density 155
$Ba(Sr_{0.5}Os_{0.5})O_3$
  crystallography 35
  density 155
  Madelung constant 41
$Ba(Sr_{0.5}Re_{0.5})O_3$
  crystallography 35
  density 155
  Madelung constant 41
$Ba(Sr_{0.5}U_{0.5})O_3$
  crystallography 35
  density 155
$Ba(Sr_{0.5}W_{0.5})O_3$
  crystallography 16, 35
  density 155
  phase transition 16

$Ba(Zn_{0.5}Os_{0.5})O_3$
  crystallography 35
  density 155
$Ba(Zn_{0.5}Re_{0.5})O_3$
  crystallography 36
  density 155
$Ba(Zn_{0.5}U_{0.5})O_3$
  crystallography 36
  density 155
$Ba(Zn_{0.5}W_{0.5})O_3$
  crystallography 36
  density 155
$Ca(Ca_{0.5}Os_{0.5})O_3$
  crystallography 36
  density 155
$Ca(Ca_{0.5}Re_{0.5})O_3$
  crystallography 36
  density 155
$Ca(Ca_{0.5}W_{0.5})O_3$
  crystallography 36
  density 155
$Ca(Cd_{0.5}Re_{0.5})O_3$
  crystallography 36
  density 155
$Ca(Co_{0.5}Os_{0.5})O_3$
  crystallography 36
  density 155
$Ca(Co_{0.5}Re_{0.5})O_3$
  crystallography 36
  density 155
$Ca(Fe_{0.5}Re_{0.5})O_3$
  crystallography 36
  density 155
$Ca(Mg_{0.5}Re_{0.5})O_3$
  crystallography 36
  density 155
$Ca(Mg_{0.5}W_{0.5})O_3$
  crystallography 36
  density 155
$Ca(Mn_{0.5}Re_{0.5})O_3$
  crystallography 36
  density 155
$Ca(Ni_{0.5}Re_{0.5})O_3$
  crystallography 36
  density 155
$Ca(Sr_{0.5}W_{0.5})O_3$
  crystallography 36
  density 155
$Pb(Ca_{0.5}W_{0.5})O_3$
  crystallography 36
  density 155
$Pb(Cd_{0.5}W_{0.5})O_3$
  crystallography 36
  density 156

$Pb(Co_{0.5}W_{0.5})O_3$
  crystallography 36
  density 156
  ferroelectricity 104
  phase transition 119
$Pb(Mg_{0.5}Te_{0.5})O_3$
  crystallography 36
  density 156
  phase transition 120
  preparation, single crystal 176
$Pb(Mg_{0.5}W_{0.5})O_3$
  crystallography 36
  density 156
  dielectric constant 100
  ferroelectricity 100
$Sr(Ca_{0.5}Mo_{0.5})O_3$
  crystallography 36
  density 156
$Sr(Ca_{0.5}Os_{0.5})O_3$
  crystallography 36
  density 156
$Sr(Ca_{0.5}Re_{0.5})O_3$
  crystallography 36
  density 156
$Sr(Ca_{0.5}U_{0.5})O_3$
  crystallography 36
  density 156
$Sr(Ca_{0.5}W_{0.5})O_3$
  crystallography 36
  density 156
$Sr(Cd_{0.5}Re_{0.5})O_3$
  crystallography 36
  density 156
$Sr(Cd_{0.5}U_{0.5})O_3$
  crystallography 36
  density 156
$Sr(Co_{0.5}Mo_{0.5})O_3$
  crystallography 36
  density 156
  phase transition 119
$Sr(Co_{0.5}Os_{0.5})O_3$
  crystallography 36
  density 156
$Sr(Co_{0.5}Re_{0.5})O_3$
  crystallography 36
  density 156
  Madelung constant 41
$Sr(Co_{0.5}U_{0.5})O_3$
  crystallography 36
  density 156
$Sr(Co_{0.5}W_{0.5})O_3$
  crystallography 37

INDEX

$Sr(Co_{0.5}W_{0.5})O_3$ (cont.)
  density 156
  ferromagnetism 127
  phase transition 119
$Sr(Cr_{0.5}U_{0.5})O_3$
  crystallography 37
  density 156
$Sr(Cu_{0.5}W_{0.5})O_3$
  crystallography 37
  density 156
$Sr(Fe_{0.5}Os_{0.5})O_3$
  crystallography 37
  density 156
  Madelung constant 41
$Sr(Fe_{0.5}Re_{0.5})O_3$
  crystallography 37
  density 156
  Madelung constant 41
$Sr(Fe_{0.5}U_{0.5})O_3$
  crystallography 37
  density 156
$Sr(Fe_{0.5}W_{0.5})O_3$
  crystallography 37
  density 156
  ferromagnetism 126
$Sr(Mg_{0.5}Mo_{0.5})O_3$
  crystallography 37
  density 156
$Sr(Mg_{0.5}Os_{0.5})O_3$
  crystallography 37
  density 156
$Sr(Mg_{0.5}Re_{0.5})O_3$
  crystallography 37
  density 156
$Sr(Mg_{0.5}Te_{0.5})O_3$
  crystallography 37
  density 156
$Sr(Mg_{0.5}U_{0.5})O_3$
  crystallography 37
  density 156
$Sr(Mg_{0.5}W_{0.5})O_3$
  crystallography 37
  density 156
$Sr(Mn_{0.5}Re_{0.5})O_3$
  crystallography 37
  density 156
$Sr(Mn_{0.5}U_{0.5})O_3$
  crystallography 37
  density 156
$Sr(Mn_{0.5}W_{0.5})O_3$
  crystallography 37
  density 156
  ferromagnetism 126

$Sr(Ni_{0.5}Mo_{0.5})O_3$
  crystallography 37
  density 156
  phase transition 119
$Sr(Ni_{0.5}Re_{0.5})O_3$
  crystallography 37
  density 156
$Sr(Ni_{0.5}U_{0.5})O_3$
  crystallography 37
  density 156
  Madelung constant 41
$Sr(Ni_{0.5}W_{0.5})O_3$
  crystallography 37
  density 156
  ferromagnetism 126
  phase transition 119
$Sr(Pb_{0.5}Mo_{0.5})O_3$
  crystallography 37
  density 156
$Sr(Sr_{0.5}Os_{0.5})O_3$
  crystallography 37
  density 156
  Madelung constant 41
$Sr(Sr_{0.5}Re_{0.5})O_3$
  crystallography 37
  density 156
  Madelung constant 41
$Sr(Sr_{0.5}U_{0.5})O_3$
  crystallography 37
  density 156
$Sr(Sr_{0.5}W_{0.5})O_3$
  crystallography 37
  density 156
$Sr(Zn_{0.5}Mo_{0.5})O_3$
  phase transition 119
$Sr(Zn_{0.5}Re_{0.5})O_3$
  crystallography 37
  density 156
  Madelung constant 41
$Sr(Zn_{0.5}W_{0.5})O_3$
  crystallography 37
  density 157
  phase transition 119

$A^{2+}(B_{0.5}^{1+}B_{0.5}^{7+})O_3$

$Ba(Ag_{0.5}I_{0.5})O_3$
  crystallography 38
  density 157
$Ba(Li_{0.5}Os_{0.5})O_3$
  crystallography 16, 38

  density 157
$Ba(Li_{0.5}Re_{0.5})O_3$
  crystallography 38
  density 157
  Madelung constant 40
$Ba(Na_{0.5}I_{0.5})O_3$
  crystallography 38
  density 157
$Ba(Na_{0.5}Os_{0.5})O_3$
  crystallography 16, 38
  density 157
$Ba(Na_{0.5}Re_{0.5})O_3$
  crystallography 38
  density 157
$Ca(Li_{0.5}Os_{0.5})O_3$
  crystallography 38
  density 157
$Ca(Li_{0.5}Re_{0.5})O_3$
  crystallography 38
  density 157
$Sr(Li_{0.5}Os_{0.5})O_3$
  crystallography 38
  density 157
$Sr(Li_{0.5}Re_{0.5})O_3$
  crystallography 38
  density 157
$Sr(Na_{0.5}Os_{0.5})O_3$
  crystallography 38
  density 157
$Sr(Na_{0.5}Re_{0.5})O_3$
  crystallography 38
  density 157
  preparation, powder 160

$A^{3+}(B_{0.5}^{2+}B_{0.5}^{4+})O_3$

$La(Co_{0.5}Ir_{0.5})O_3$
  crystallography 38
  density 157
$La(Cu_{0.5}Ir_{0.5})O_3$
  crystallography 38
  density 157
$La(Mg_{0.5}Ge_{0.5})O_3$
  crystallography 38
  density 157
$La(Mg_{0.5}Ir_{0.5})O_3$
  crystallography 38
  density 157
$La(Mg_{0.5}Nb_{0.5})O_3$
  crystallography 38
  density 157

INDEX

$La(Mg_{0.5}Ru_{0.5})O_3$
  crystallography 38
  density 157
  electrical conductivity 66
  preparation, single crystal 178
  X-ray diffraction 56
$La(Mg_{0.5}Ti_{0.5})O_3$
  crystallography 38
  density 157
$La(Mn_{0.5}Ir_{0.5})O_3$
  crystallography 38
  density 157
$La(Mn_{0.5}Ru_{0.5})O_3$
  crystallography 38
  density 157
$La(Ni_{0.5}Ir_{0.5})O_3$
  crystallography 38
  density 157
  electrical conductivity 66
  preparation, single crystal 178
$La(Ni_{0.5}Ru_{0.5})O_3$
  crystallography 38
  density 157
  electrical conductivity 66
  preparation, single crystal 178
$La(Ni_{0.5}Ti_{0.5})O_3$
  crystallography 38
  density 157
$La(Zn_{0.5}Ru_{0.5})O_3$
  crystallography 38
  density 157
  electrical conductivity 66
  preparation, single crystal 178
$Nd(Mg_{0.5}Ti_{0.5})O_3$
  crystallography 38
  density 157

$A^{2+}(B^{1+}_{0.25}B^{5+}_{0.75})O_3$

$Ba(Na_{0.25}Ta_{0.75})O_3$
  crystallography 17, 39
  density 157
$Sr(Na_{0.25}Ta_{0.75})O_3$
  crystallography 17, 39
  density 157

$A^{2+}(B^{3+}_{0.5}B^{4+}_{0.5})O_{2.75}$

$Ba(In_{0.5}U_{0.5})O_{2.75}$
  crystallography 39
  density 157

$A^{2+}(B^{2+}_{0.5}B^{5+}_{0.5})O_{2.75}$

$Ba(Ba_{0.5}Ta_{0.5})O_{2.75}$
  crystallography 39
  density 157
$Ba(Fe_{0.5}Mo_{0.5})O_{2.75}$
  crystallography 39
  density 157
$Sr(Sr_{0.5}Ta_{0.5})O_{2.75}$
  crystallography 39
  density 157

CARBIDES

$AlFe_3C$
  crystallography 185
$AlMn_3C$
  crystallography 185
  ferromagnetism 187
Fe_3SnC
  crystallography 185
$GaMn_3C$
  crystallography 185
Mn_3ZnC
  crystallography 185
  electrical conductivity 187
  ferromagnetism 187

HALIDES

$CsCaF_3$
  crystallography 185
$CsCdBr_3$
  crystallography 185
$CsCdCl_3$
  crystallography 185
$CsFeF_3$
  crystallography 185
  preparation, single crystal 184
$CsGeCl_3$
  crystallography 185
$CsHgBr_3$
  crystallography 185
$CsHgCl_3$
  crystallography 185
$CsMgF_3$
  crystallography 185
$CsPbBr_3$
  crystallography 185
$CsPbCl_3$
  crystallography 185
$CsZnF_3$
  crystallography 185
$KCaF_3$
  crystallography 185
$KCdF_3$
  crystallography 185

INDEX 207

KCoF$_3$
  crystallography 185
KCrF$_3$
  crystallography 185
KCuF$_3$
  crystallography 185
  ferromagnetism 188
  preparation 183
KFeF$_3$
  crystallography 185
  ferromagnetism 188
  preparation 183
KMgF$_3$
  crystallography 185
KMnF$_3$
  crystallography 185
  ferromagnetism 188
  preparation 183
KNiF$_3$
  crystallography 185
  ferromagnetism 188
  preparation 183
KZnF$_3$
  crystallography 185
LiBaF$_3$
  crystallography 185

NaZnF$_3$
  crystallography 185
RbCaF$_3$
  crystallography 185
RbCoF$_3$
  crystallography 185
RbFeF$_3$
  crystallography 185
  ferromagnetism 188
  preparation, single crystal 184
RbMgF$_3$
  crystallography 185
RbMnF$_3$
  crystallography 186
  ferromagnetism 188
RbZnF$_3$
  crystallography 186
K(Cr$_{0.5}$Na$_{0.5}$)O$_3$
  crystallography 186
  preparation 182
K(Fe$_{0.5}$Na$_{0.5}$)O$_3$
  crystallography 186
  preparation 182
K(Ga$_{0.5}$Na$_{0.5}$)O$_3$
  crystallography 186
  preparation 182

HYDRIDES

LiBaH$_3$
  crystallography 186
  preparation 184
LiEuH$_3$
  crystallography 186
LiSrH$_3$
  crystallography 186
  preparation 184

NITRIDES

Fe$_4$N
  crystallography 186
  ferromagnetism 188
  preparation 184
Mn$_4$N
  crystallography 186
  ferromagnetism 188
  preparation 184

Fe$_3$NiN
  crystallography 186
  ferromagnetism 188
  preparation 184
Fe$_3$PtN
  crystallography 186
  ferromagnetism 188
  preparation 184

OTHER TITLES IN THE
SERIES IN SOLID STATE PHYSICS

Vol. 1 F. P. JONA & G. SHIRANE—Ferroelectric Crystals
Vol. 2 J. H. SCHULMAN & W. D. COMPTON—Colour Centers in Solids
Vol. 3 J. FRIEDEL—Dislocations
Vol. 4 S. V. VONSOVSKII—Ferromagnetic Resonance